T0135751

Lamb Waves for Structural Health Monitoring in Viscoelastic Composite Materials

Vom Fachbereich Produktionstechnik

der

UNIVERSITÄT BREMEN

zur Erlangung des Grades

Doktor-Ingenieur

genehmigte

Dissertation

von

Dipl.-Ing. Mircea Calomfirescu

Gutachter: Prof. Dr.-Ing. Axel S. Herrmann
             Prof. Dr.-Ing. Victor Giurgiutiu (University of South Carolina, USA)

Tag der mündlichen Prüfung: 28. April 2008

Science-Report aus dem Faserinstitut Bremen

Hrsg.: Prof. Dr.-Ing. Axel S. Herrmann

ISSN 1611-3861

Bibliografische Information der Deutschen Nationalbibliothek

Die Deutsche Nationalbibliothek verzeichnet diese Publikation in der
Deutschen Nationalbibliografie; detaillierte bibliografische Daten sind
im Internet über http://dnb.d-nb.de abrufbar.

ISBN 978-3-8325-1946-9

Logos Verlag Berlin GmbH
Comeniushof, Gubener Str. 47,
10243 Berlin
Tel.: +49 030 42 85 10 90
Fax: +49 030 42 85 10 92
INTERNET: http://www.logos-verlag.de

ACKNOWLEDGEMENTS

This Ph.D thesis arose during my employment as a scientific research assistant at the Faserinstitut Bremen e.V. at the University of Bremen.

Professor Dr.-Ing. Axel S. Herrmann gave me the opportunity to intensely work on the issues of this thesis. I am forever grateful for his outstanding support, his excellent ideas and the scientific freedom which made this research possible.

Special thanks are extended to Professor Dr.-Ing. Victor Giurgiutiu from the Laboratory of Active Materials and Smart Structures at the University of South Carolina and the Air Force Research Laboratory for the opportunity to work in his lab and for many valuable discussions. He reviewed this thesis and was part of the dissertation committee.

Thanks to all of my colleagues at the Faserinstitut, especially Mr. Dipl.-Ing Holger Purol, the head of the Department FVSV I was part of.

Lastly, I wish to thank my parents and my wonderful wife Diana for their support and confidence in me.

The research presented within this thesis was funded by the "Deutsche Forschungsgemeinschaft (DFG)" for which I am thoroughly appreciative.

Mircea Calomfirescu
Bremen, December 2007

ABSTRACT

Structural Health Monitoring (SHM) is a novel philosophy for an autonomous, built-in nondestructive evaluation of structural "health" on demand to reduce life-cycle costs, increase safety and reduce structural weight. This dissertation investigates ultrasonic guided waves, particularly Lamb waves, and their propagation properties as a method to perform Health Monitoring of viscoelastic composite structures.

One of the objectives of this work lies in the analytical description of Lamb wave propagation in anisotropic, viscoelastic composite materials by a higher order plate theory taking into account traversal shear and rotational inertia. To take account of material damping the theory of complex moduli was adopted. For the prediction of dispersion and attenuation behavior of guided waves, the developed models based on an analytical, higher order plate theory were implemented into a software enabling the calculation of S_0, A_0 and SH_0 modes, which are the most commonly used guided wave types for SHM applications. The results were verified successfully experimentally. For experimental verification, an experimental test set-up was built up in which Lamb waves could be excited and their propagation was measured by small and lightweight, surface-bonded piezoelectric wave active sensors (PWAS). The frequency range was $10 - 500$ kHz. Moreover, the dispersion and attenuation behavior of Lamb wave propagation was experimentally and analytically studied for six carbon fiber reinforced plastic (CFRP) laminates, which are typical in the aerospace industry.

Based on these fundamental studies, different SHM applications were analyzed in detail: on the one hand, a method for impact detection, localization and quantification; on the other hand, a novel method for the non-destructive, quantitative viscoelastic material characterization was proposed and developed.

In order to detect and quantify impact events, an existing method was extended and further developed for application to anisotropic, viscoelastic composites. The flexural waves excited by the impact were measured by surface bonded PWAS and the signals were analyzed by the Wavelet transform (WT), which increased significantly the accuracy of the arrival time detection. The impact was applied with an instrumented, hand-held impuls hammer. For the reconstruction of the impact force history, an analytical structural model was adopted, which calculates the propagation of flexural waves due to the transverse impact. The analytical model was extended to account for material damping. A verification of the proposed methodology showed good agreement with the real impact position and force-time history.

Another important issue in this thesis is the development of a novel non-destructive, quantitative viscoelastic material characterization method based on Lamb wave dispersion and attenuation measurements using PWAS. The inversion of the viscoelastic material stiffness coefficients was realized by a non-linear linear simplex optimization method. The results showed good agreement with the material properties measured by conventional (destructive) tensile tests and by another state-of-the art, non-destructive technique. The method developed in this dissertation is appropriate for in-situ applications.

KURZFASSUNG

Structural Health Monitoring (SHM) bezeichnet eine neuartige Philosophie zur automatischen Strukturzustandüberwachung „auf Abruf" mit dem Ziel Betriebskosten zu senken, die Sicherheit zu erhöhen und Strukturgewicht einzusparen. Im Rahmen dieser Dissertation werden so genannte geführte Wellen (oder Lamb Wellen) und ihre Ausbreitungseigenschaften für unterschiedliche Structural Health Monitoring Anwendungen an viskoelastischen Faserverbundstrukturen untersucht.

Einer der Schwerpunkte der Arbeit liegt in der analytischen Beschreibung der Lamb Wellen Ausbreitung in anisotropen, viskoelastischen Faserverbundwerkstoffen mit Hilfe einer Plattentheorie höherer Ordnung unter Berücksichtigung von transversalem Schub und Rotationsträgheiten. Zur Berücksichtigung der anisotropen Materialdämpfung des Faserverbundes, ein in der heutigen Literatur sehr vernachlässigtes Thema, wird die Methode der komplexen Moduln verwendet. Zur Berechnung des Dispersions- und Dämpfungsverhaltens der Lamb Wellen Modi S_0, A_0 und SH_0, die die am meisten verwendeten Modi im SHM Bereich darstellen, wurden die entwickelten Methoden in eine Software implementiert. Die Ergebnisse wurden experimentell erfolgreich überprüft. Dazu wurde ein experimenteller Prüfstand aufgebaut, an dem mit Hilfe von Piezowandlern (so genannter „Piezoelectric Waver Active Sensors" (PWAS)), Lamb Wellen angeregt und aufgenommen wurden. Dabei wurde ein Frequenzbereich von 10 – 500 KHz untersucht. Die Übereinstimmung zwischen dem analytischen Modell und den experimentellen Ergebnissen war sehr gut. Das Dämpfungsverhalten von Lamb Wellen wurde für sechs luftfahrttypische, kohlenstofffaserverstärkte Laminate zudem experimentell und analytisch untersucht.

Ausgehend von diesen Grundlagenuntersuchungen wurden unterschiedliche SHM-Anwendungen näher untersucht: zum einen die Detektion und Bewertung von Impactvorgängen und zum anderen die zerstörungsfreie, quantitative Bestimmung viskoelastischer Materialeigenschaften.

Zur Detektion und Bewertung von Impactvorgängen wurde eine bereits zuvor entwickelte Methode für die Anwendung auf Faserverbundstrukturen weiterentwickelt und implementiert. Die mit Hilfe von PWAS aufgenommen Signale nach einem Impact werden dabei mit der Wavelet- Transformation analysiert. Dies erleichtert die Bestimmung von Wellenankunftszeiten und macht die Ortung von Impacts in Faserverbundstrukturen zuverlässiger. Zur Aufnahme des Stoßkraftverlaufs, wurden die Impacts mit einem Impulshammer aufgebracht. Zur Rekonstruktion des Stoßkraftverlaufes und damit Bewertung des Impacts wurde ein analytisches Strukturmodell verwendet, das die Ausbreitung von Biegewellen nach einem Impact beschreibt. In diesem Strukturmodell wurde erstmals die anisotrope Materialdämpfung von Faserverbundwerkstoffen berücksichtigt. Die entwickelte Methode wurde an kohlenstofffaserverstärkten Laminaten erfolgreich verifiziert. Dabei konnte die Impactstelle bis auf wenige cm Genauigkeit lokalisiert werden und der Stoßkraftverlauf mit ausreichender Genauigkeit rekonstruiert werden.

Ein weiterer Schwerpunkt dieser Arbeit liegt in der Entwicklung einer neuartigen, Lamb Wellen basierten Methode zur Bestimmung der viskoelastischen Materialeigenschaften von Faserverbundlaminaten ausgehend vom Dispersions- und Dämpfungsverhalten der Plattenwellen. Im Grunde stellt die Methode das inverse Problem des zuvor entwickelten Modells zur Berechnung von Lamb Wellen dar. Dabei sind jedoch in diesem Fall nicht die Materialeigenschaften bekannt und das Lamb Wellen Ausbreitungsverhalten gesucht, sondern die Materialeigenschaften, zu einem bereits bekannten Lamb Wellen Verhalten, sind die gesuchten Größen. Diese Methode wurde mit Hilfe eines nichtlinearen Simplex Optimierungsverfahrens realisiert. Die Ergebnisse dieser neuartigen, im Rahmen dieser Arbeit entwickelten Methode, wurden Materialkennwerten aus zerstörenden und zerstörungsfreien Untersuchungen, die dem heutigen Stand der Technik entsprechen, erfolgreich gegenübergestellt. Die in dieser Dissertation entwickelte Methode ist im Gegensatz zu allen heute verfügbaren Techniken, für „in-situ" Anwendungen geeignet und kann leicht automatisiert werden. Dadurch ließen sich die Materialkennwerte einer sich im Betrieb befindenden Struktur bestimmen.

Table of Content

1   Introduction

The intention of this first chapter is to introduce the topic of Structural Health Monitoring (SHM) and to define the role of this thesis in the SHM framework. The background and the state of the art, including some perspectives for the next 20 years, are given. The objective, the content, and the contribution of this research work to the advancement of the state of the art will be pointed out in the chapters 1.3 and 1.4.

1.1   Background of Structural Health Monitoring

Structures in all engineering areas are subjected to damage resulting from overloading, fatigue, unexpected events, such as impacts, environmental effects, including corrosion and many other unpredictable events. Particularly in aerospace and civil engineering, failures of the primary structure may put at risk human life, as numerous aircraft and bridge accidents have shown in a dramatic way in the past. For example the three aircraft crashes of the Comet, in 1953 and 1954, due to metal fatigue and damage tolerance design faults [Com07], the crash of the Aloha Airlines Boeing 737-200 in 1988, which was also caused by metal fatigue exacerbated by crevice corrosion (the plane operated in a salt water environment) [Wik07], the more recent crush of the China Airlines Boeing 747-200B in 2002 (also the result of metal fatigue, in that case due to inadequate maintenance after a previous incident in 1980 [Wik07a]). All of these accidents demonstrated in a dramatic way the importance of inspection and maintenance as a safety issue.

Many different inspection and non-destructive testing (NDT) techniques are available and applied in the aerospace industry for many decades, most of them being developed in the 1960s. These conventional NDT techniques range from basic visual inspection, ultrasonic inspection and thermography to radiography, shearography and eddy-current. Most of the conventional NDT- techniques are time-consuming processes that require meticulous scans over large areas of the structure. Moreover, the results of such inspections and their reliability are often strongly influenced by the human factor (e.g. the experience and attention of the inspector). Despite all the accidents, the economic factor, which is related to high maintenance and inspection costs, but also to time the structure/aircraft is out of service, represents a significant, crucial factor for the end-user, who has to decrease his direct operational costs (DOC). For commercial and military aircraft, it is estimated that up to 27% of the average life cycle costs are related to inspection and repair [Hal99].

Motivated by these aspects, the last decade has shown a range of novel NDT techniques, for which the term "Structural Health Monitoring" (SHM) has been introduced. The main difference between SHM and the conventional NDT techniques is that SHM utilizes (permanently) integrated sensors, which provide continuous or on-demand information about the state at critical high-stress locations of safety-critical structures. The potential of SHM arise from a number of elements such as: reduced inspection and maintenance costs, improved performance and safety, reduced out of service times and, last but not least, reduction of the influence of the human factor. Furthermore, as recently highlighted by Schmidt [Sch07], other possible benefits of SHM systems are related to structural weight reductions, if structural health monitoring could be implicated in the structural design process (Fig. 1.1). Schmidt [Sch07] has shown that a mass reduction potential up to 20% is possible, if SHM could be implicated in the structural design.

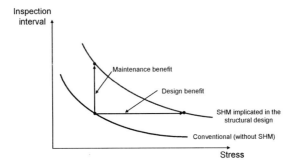

Fig. 1.1: Maintenance and design benefits due to SHM, after [Sch07]

The concept of Structural Health Monitoring can be compared to that of the human nervous system (Fig. 1.2), which also benefits from many different "sensors", which transmit signals (e.g. pain) to the main "computer", the brain.

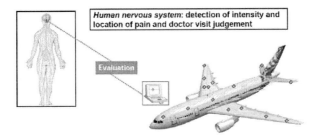

Fig. 1.2: Concept of Structural Health Monitoring [Boc06]

The main components of an SHM- system can be classified into the following categories:
- Sensors/actuators
- Data acquisition
- Signal processing
- Materials and structures
- Computation
- Smart software

As mentioned before, SHM is a multidisciplinary field, in which researchers and engineers from different disciplines such as materials science, computer science, non-destructive evaluation, dynamics, mechanical engineering, etc. have to contribute and be involved. The development of structural health monitoring systems covered the following purposes:

- Load history monitoring
- Impact detection and localization
- Damage identification/localization/quantification (e.g. impact damages in composites, corrosion, fatigue cracking, disbond in stiffened composite panels, manufacturing damages in composites)
- Residual stress monitoring
- Residual stiffness and strength monitoring

For the identification of damages, different SHM approaches have been proposed in the past, which can be classified as either global or local. Global approaches are based upon vibration measurements of the structure in the lower frequency range (< 50 kHz), for which active excitation is not required. Such global methods are typically only sensitive to fairly large levels of damages and may be used therefore to monitor large areas for locating suspect positions that may then be covered in detail by a further inspection technique. During the last years, local diagnostic methods have generated considerable interest in the SHM community. These methods typically consider high frequencies, mainly within the range of 10 kHz to 1 MHz. In this context the most widely discussed approach is to utilize ultrasonic Lamb waves, which interact with damages. The development of small, inexpensive and lightweight piezoelectric elements, which can be utilized both as actuators and as sensors, is a further key point in the development of local Lamb wave based SHM techniques.

1.2   State of the Art and Perspectives

SHM is still a relatively young field of research, not older than about 15 to 20 years starting in the early 1990s. However, it is difficult to fix the date of the birth of SHM, since some aspects have already been developed and presented much earlier by using different terms such as: (in-situ) non destructive evaluation (NDE), structural control, smart materials and many other terms.

In 1997, the first international conference devoted only to SHM was initiated by Fu-Kuo Chang at the Stanford University. Since then, this International Workshop on Structural Health Monitoring (IWSHM) is being held every other year (1997, 1999, 2001...etc.). Since 2002 every other year (alternatively with the aforementioned Stanford Workshop), the European Workshop on SHM (EWSHM) takes place at different locations in Europe (Cachan (2002), Munich (2004), Granada (2006)). Examining the number of papers presented at these workshops, one observes a constant progression from 65 papers at the 1st IWSHM in 1997 to more than 1000 papers ten years later at the 6th IWSHM in 2007.

A comparable evolution to that of the SHM conferences has occurred in scientific journals related to the topic of SHM. The International Journal of Structural Health Monitoring was launched in 2002. It was published twice in 2002 and is issued quarterly since 2003. Other journals that cover SHM topics are "The Journal of Intelligent Material Systems and Structures", "The Smart Materials and Structures Journal", "The Shock and Vibration Digest" and "The Journal of Sound and Vibration".

As mentioned above, there are many different SHM techniques and approaches; however, the intention of the review given here is focused on techniques applicable on composites materials for local SHM evaluation. For a more comprehensive literature review, see reference [Soh03].

One method for local SHM discussed in the literature is the impedance method. This method has been proposed by Liang [Lia96] and Rogers [Rog97] and further developed by Giurgiutiu [Giu00] and co-workers and Inman and co-workers [Par03]. The basic requirement for this method is a piezoelectric element, which is able to work as actuator and as sensor. Most researchers applied the impedance method in the higher frequency range, typically >30 kHz. The basic idea of this method is that a damage in the structure changes the structural impedance and the resulting electrical impedance spectrum of the sensor mounted on the structure. The impedance method has been referred to as a very local, qualitative method. In reference [Par03] it is mentioned that a sensing radius of 0.4 m can be reached in composite structures and up to 2 m in a metal beam for damage detection.

Another SHM method is to utilize elastic waves and their propagation to analyze mechanical properties and structural damage. Various types of methods based on sound and ultrasound have been developed. In general, all wave based SHM approached can be classified into passive and active approaches. As illustrated in Fig. 1.3, passive approaches need only sensors to "listen" to the structure, but no actuators to excite the structure are required.

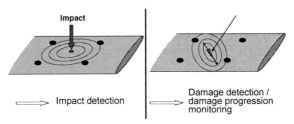

Fig. 1.3: Passive, wave based SHM - approach

The passive approach can be adopted for impact detection, for example. The basic idea in this case is to identify impact events from transient strain measurements at discrete points of the structure subjected to impact loadings. In the literature there are two basic approaches discussed for impact detection: model-based techniques [Doy87], [Sey01], [Par05] and neural networks [Wor00], [Hay01]. Model-based techniques are based on mathematical models which describe the structural behaviour. Neural networks do not need an understanding of the mechanical problem, since they calculate outputs through neural networks algorithms which are trained by a set of experimental data. The passive approach has also been adopted for damage detection and damage progression, by just "listening" to the acoustic signals generated by the damage itself. This method is generally referred to as acoustic emission (AE).

Another technique adopted by many researchers is the active approach. Here not only sensors but also actuators are required to excite the structure. The excitation can be collected either at the same position, this is typically called "pulse-echo method" or at different positions using the "pitch-catch method" (Fig. 1.4).

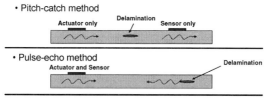

Fig. 1.4: Active SHM - approaches using Lamb waves

Nowadays, the most widely discussed method for SHM damage detection is to utilize Lamb waves. Lamb waves are a type of ultrasonic waves, which travel in thin-walled structures and are dispersive, i.e. the velocity of these waves depends on the frequency. These waves were first theoretically explained and derived by Horace Lamb in 1917 [Lam17]. One of the first researchers who proposed to utilize this type of waves for damage detection was Worlton [Wor57] in the year 1957. However, the method proposed by Worlton required mobile probes, which could not be integrated or built into the structure. Further research on the application of Lamb waves and their interaction with damages was performed much later at the Imperial College by Alleyne and Cawley [All92], [All92a]. Later research works at the Imperial College were focussed on the influence of the dispersion of Lamb waves on their propagation characteristics. Wilcox, Lowe and Cawley developed analytical methods using narrow band, tone-burst signals [Wil01].

The automatic monitoring of the structural health using Lamb waves and integrated piezoelectric wafer active sensors (PWAS) was pioneered by Chang [Cha95], [Cha98] and his group at the Stanford University. Chang also used narrow band, tone-burst excitation signals in order to minimize the influence of dispersion. The method described by Chang compared so called baseline Lamb wave measurements of a pristine structure with measurements of damaged structure. Both measurements were compared, the basic idea being that the wave propagation characteristics change if a damage is present.

The most widely used sensors for SHM applications are piezoelectric transducers. Their big advantage is that they can be used as sensors as well as actuators. Different piezoelectric materials are available, however, the most widely used are the ceramic lead zirconate titanate (PZT) and the polymer polyvinylidene fluoride (PVDF). Important developments in this area include piezoceramic paints [Egu93], Smart Layer[TM] [Cha98a] and piezoceramic fibres [Yos95]. Other sensor technologies often discussed for SHM applications are optical fibre sensors [Glo90], [Udd92], [Mea01] and Micro-Electro-Mechanical Systems (MEMS) [Khu00]. Optical fiber sensors can be used for strain and temperature measurements and have the advantage that they are immune to electromagnetic fields.

Many prototypes and laboratory demonstrations of SHM methods and systems have already been presented and displayed at meetings, conferences and workshops (for example during the IWSHM in 2005 and 2007 and at the EWSHM in 2006). However, although significant advances in developing and maturing SHM have been made (especially during the last ten years), SHM (either local, global or combined) has not really managed to transit from research to practice in the aerospace industry. One of the reasons mentioned by Achenbach [Ach07] may be the strong competition from the existing NDI infrastructure for scheduled inspection with a good safety record and accepted prognosis methods. SHM has still to demonstrate its economic benefit with a high probability of detection (POD) and with near zero false alarm rates. Chang [Cha06] mentioned the need of testing facilities for assessing the integrated structures and appropriate design methodologies and validation procedures for incorporating the technologies into the design of intelligent structures with the capability of self-diagnostics and monitoring of practical applications. Environmental effects are not yet enough analyzed and quantified in field studies. Fig. 1.5 illustrates the vision of the SHM development for the next 15 years recently given by Meyendorf [Mey06]. In Meyendorf's vision the next years have to manage the jump from laboratory prototypes into practice, first for local monitoring. Basic sensor principles are almost fully developed, but there will be much more work in developing smart sensors and networks during the next 5 to 10 years. According to Meyendorf [Mey06], the main goal is the development of a health management and control system might follow three main paths: (1) lab prototypes (today), (2) local monitoring in practice (in the next 8 to 10 years) and (3) global monitoring.

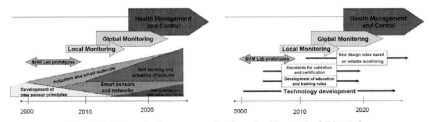

Fig. 1.5: Attempt of a prognosis given by Meyendorf [Mey06]

1.3   Objective of the Thesis

The objective of this thesis is to make a contribution to enhance the effectiveness of using Lamb waves for different Structural Health Monitoring (SHM) applications in composite materials. In comparison to metallic materials such as aluminum and steel, composites possess much higher material damping properties, which lead to a higher Lamb wave attenuation. The attenuation in composites is anisotropic and it depends on frequency and Lamb wave mode. The fundamental understanding as well as the quantification of the attenuation in composites is a crucial factor for the development and optimization of sensor networks for SHM application. The main problem is to decide how far Lamb waves can travel and can be still distinguished from noise (Fig. 1.6).

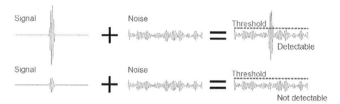

Fig. 1.6: Lamb wave detectability [Wil06]

Therefore, one of the main objectives of this research is to develop a mathematical model for the prediction of Lamb wave dispersion and attenuation behavior in composites of arbitrary lay-ups. The effectiveness of the developed model will be demonstrated by implementing it into a software and the results will be compared to experiments. Based on this approach, the further objectives of this research are to develop and demonstrate applications. In this context, an existing method for impact location and contact force reconstruction will be extended and further developed for the application on composites. Lastly, an inverse solution of the mathematical model for the prediction of Lamb waves will be developed which enables the in-situ, quantitative composite materials characterization based on velocity and attenuation dispersion data.

The main contributions of this dissertation to the advancement of the state of the art are listed as following:

- Development and implementation of a higher order plate theory for viscoelastic Lamb wave propagation in anisotropic materials
- Experimental demonstration of the applicability of the higher order plate theory for the prediction of anisotropic Lamb wave attenuation in composites
- Providing extensive experimental Lamb wave attenuation measurements on different CFRP laminates with PWAS transducers for the validation of the proposed Lamb wave propagation model
- Development of a novel quantitative inverse-solution for material characterization in order to obtain viscoelastic material properties from experimental Lamb wave velocity and attenuation dispersion curves
- Extension and further development of an existing method for impact detection, location and contact force reconstruction for application to composites

1.4   Content of the Thesis

Chapter 2 provides the theoretical background on Lamb waves and highlights different wave attenuation mechanisms, pointing out the attenuation mechanism due to material damping, which is the focus of this research. Some insight in the fundamentals of piezoelectric materials and piezoelectric waver active sensors (PWAS) are given in a section of this chapter. PWAS are used in this research for sensing and actuating of Lamb waves during experimental tests. A crucial factor of experimental testing consists in signal processing for extracting useful information. Therefore, the main signal processing techniques used in this research are also described in chapter 2. Since the propagation of Lamb waves in composites was not only studied experimentally and analytically, but also analyzed in this research by numerical methods, such as explicit finite element analysis (E-FEA), an introduction of E-FEA is given as well in one of the sections of chapter 2.

Chapter 3 presents the derivation of a higher order plate theory with the inclusion of material damping. The classical plate theory is not applicable for Lamb waves, in particular at higher frequencies, since it neglects transverse shear and rotational inertia. A comparison of different methods for the calculation of Lamb waves is given, pointing out their advantages and limitations. Material models for accounting of viscoelastic material damping in composites are presented.

Chapter 4   presents the implementation of the viscoelastic higher order plate theory model and introduces the software developed in this research. A short discussion of the results closes this chapter.

Chapter 5 gives some insight of a non destructive material characterization method, which was adopted in this research for the measurement of the viscoelastic stiffness coefficients of the carbon fiber reinforced plastic (CFRP) laminates. A short discussion and a summary of the obtained results are given at the end of this chapter, demonstrating the high anisotropy of CFRP laminates damping.

As mentioned before, the mathematical models for Lamb wave propagation were validated and compared to experimental methods. The experimental method and the test set-up for the measurement of Lamb wave dispersion and attenuation curves are presented in Chapter 6. Hereby, Lamb waves are excited and collected by PWAS transducers in the pitch-catch mode. The excitation signal and the excitability of Lamb waves are presented in one of the sections of this chapter. Finally, a comparison between the analytical predictions and experimental results is discussed.

Since one main objective of this dissertation was to demonstrate different SHM applications, experimental investigations for the development of an impact detection and classification approach are performed and presented in Chapter 7. An instrumented, hand held impuls hammer was used to experimentally impact the CFRP plates.

Chapter 8 studies the applicability and the accuracy of commercial Explicit Finite Element Analysis (E-FEA) codes for flexural wave propagation in the near field (where material damping is negligible). For computation, the E-FEA code LS-DYNA was used, since it offers many interesting material models including failure criteria and degradation models for composites. However, since the actual version of LS-DYNA is not able to consider anisotropic material damping, far field Lamb wave propagation studies at higher frequencies (10 – 500 kHz), for which material damping is a crucial factor, were not performed. For the short-time, transient impact loading the impact force history from experimental testing was used. The results of the E-FEA were finally compared to the results obtained experimentally with PWAS transducers.

Chapter 9 presents how an existing impact identification method was extended and further developed for application to anisotropic composites. The arrival time detection, which is a crucial point for localizing the impact events, was performed using the Wavelet Transform (WT). The localization of the impact force was performed by a non-linear optimization algorithms using velocity dispersion information provided by the higher order plate model developed in Chapter 3. The reconstruction of the impact force was also performed by an optimization method, based on an analytical structural model. In this context, material damping was implemented in the analytical structural model for prediction of flexural waves due to a point impact.

In Chapter 10, a novel approach for quantitative and non-destructive material characterization is developed and presented. The basic idea of this novel approach is to use the dispersion and attenuation Lamb wave behavior of a composite in order to determine its viscoelastic material properties. Thus, it is an inverse solution of the higher order plate theory model developed in this research. For this application, the dispersion behavior is given and the material properties are the unknown values. The inverse solution was performed by a non-linear simplex optimization algorithm. Finally, the results were compared to non-destructive tests provided by the immersion technique presented in Chapter 5 and by conventional destructive tensile tests. The agreement was very good. However, this method was only applied to unidirectional composites, but may be further developed. The composite characterization method can be applied in-situ with the same PWAS as used for different other SHM applications.

Chapter 11 summarizes the main results of this doctoral research and makes suggestions for future work that can be developed from these results.

2 Theoretical background

This chapter provides the theoretical background for Lamb waves and describes the most important attenuation mechanisms of this type of waves. The mechanical and piezoelectric relations for typical piezoelectric materials are derived and explained in detail, since thin wafers of such materials are used for sensing and actuating purposes during the experiments conducted in this research. For example, the equation for the conversion of electrical signals from sensors into mechanical strains is reviewed in this context.

Furthermore, the theoretical background of different signal processing methods, such as denoising techniques and Hilbert transformation are presented. All these techniques are necessary mathematical tools to be used during the experimental investigations.

2.1 Lamb waves

Lamb waves have been described first by Horace Lamb in 1917 [Lam17]; such waves are discussed in detail by Viktorov [Vik67], Rose [Ros99] and Achenbach [Ach99]. Lamb waves are a type of ultrasonic, dispersive waves, which are also known as guided plate waves, due to the fact that they are guided between two parallel free surfaces, the upper and the lower surfaces of the plate [Giu04].

For each frequency more than one wave mode exists. At low frequencies these wave modes are S_0 and A_0 (Fig. 2.1). Even though the guided wave mode SH_0 (Fig. 2.1) is strictly speaking not a Lamb wave mode, it is also considered in this research. The symmetrical Lamb wave modes are called, S_0, S_1, S_2,..., and the anti-symmetric ones A_0, A_1, A_2..., starting with the mode that has the lowest frequency for a given wavenumber. The particle motion of the Lamb wave modes is contained in the vertical plane. The shear horizontal (SH) modes have particle motion contained in the horizontal plane. The essence of the analysis is that standing waves are established in the transverse direction while propagating waves are developed in the lengthwise direction.

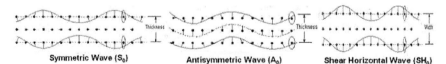

Symmetric Wave (S₀)   Antisymmetric Wave (A₀)   Shear Horizontal Wave (SH₀)

Fig. 2.1: Low order Lamb and guided wave modes [Leh06]

Lamb waves are the most widely used guided waves for damage detection [Sta04]. Moreover, Lamb waves are excited by impact events on thin walled structures.

The characteristic equations for Lamb wave propagation in isotropic, homogeneous materials, the Rayleigh-Lamb dispersion equations, can be expressed as [Vik67]

$$\frac{\tan(k_q \cdot h)}{\tan(k_p \cdot h)} = -\frac{4 \cdot k^2 \cdot k_p \cdot k_q}{\left(k_q^2 - k^2\right)^2} \tag{2.1}$$

for symmetric modes and

$$\frac{\tan(k_q \cdot h)}{\tan(k_p \cdot h)} = -\frac{\left(k_q^2 - k^2\right)^2 4 \cdot k^2 \cdot k_p \cdot k_q}{4 \cdot k^2 \cdot k_p \cdot k_q} \tag{2.2}$$

for antisymmetric modes. These equations cannot be solved analytically, therefore a numerical solution is required in order to predict velocities of different Lamb waves modes of frequency f. Fig. 2.2 illustrates a typical dispersion diagram for an aluminum plate of 2 mm thickness.

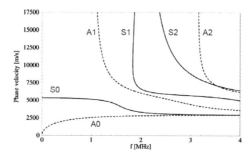

Fig. 2.2: Lamb wave dispersion characteristics [Leh06]

2.2 Wave attenuation mechanisms

Whereas the propagation of Lamb waves in elastic, isotropic materials has been investigated by many researchers, the propagation of these waves in viscoelastic, attenuative, anisotropic materials, such as composites, has been investigated by only very few e.g. [Cas00], [Nea03] and [Bar06]. However attenuation in composites is an important issue since it determines how far the different guided waves may travel. The definition of attenuation in the context of ultrasonic Lamb waves is related to the loss of wave amplitude with propagation distance. As discussed by Pollock [Pol86], there are many factors contributing to attenuation, the four main factors being:
1. geometric spreading of the wave (energy conservation, no dissipation!)
2. material damping
3. dissipation of the wave into adjacent media
4. losses related to wave dispersion.

In the near field, close to the excitation source, the geometric spreading is the dominant factor for wave attenuation; for plate-like structures and a pointwise, omnidirectional excitation source, the wave amplitude A decreases inversely with the square root of the distance of propagation. This relationship can be expressed as

$$A_2 = \sqrt{\frac{d_1}{d_2}} \cdot A_1 \qquad (2.3)$$

where A_1 and A_2 denotes the wave amplitudes and d the distance from the source.

The second factor, which is the dominating influence in the far field and the factor mostly discussed in this research, is the wave attenuation due to material damping. The material damping of composites is influenced by the components, i.e. fibers and matrix, as well as micro structural processes in the material. According to Hoffmann [Hof92] the main sources for energy dissipation and material damping in composites are:

- form, geometry, type and orientation of the fiber reinforcement
- fiber / matrix interaction, e.g. stress concentration and stress peaks at interfaces
- friction between fiber and matrix
- imperfections in the laminate (e.g. pores, microcracks, delaminations).

Thus energy can be dissipated by many different factors and extracted from the mechanical system. This type of attenuation, due to material damping, is usually approximated by an exponential relationship of attenuation with respect to distance, which can be expressed as follows:

$$A_2 = A_1 \cdot e^{-k''(d_2 - d_1)} \qquad (2.4)$$

where k'' is the attenuation factor.
Due to the high material damping of composites, attenuation leads to a decay of the amplitude of Lamb waves in composites much stronger that in metallic structures and is therefore a crucial attenuation factor for far field wave propagation in composite structures. Material damping in composites is anisotropic, showing higher damping factors in matrix dominated directions (e.g. transverse direction in an unidirectional laminate).

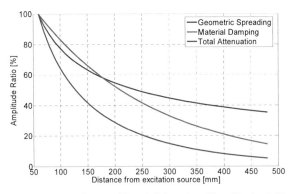

Fig. 2.3: Geometrical spreading vs. Material damping (Amplitude at 60 mm was set to 100%, attenuation factor k"= 0.04 dB/mm)

The last but one attenuation mechanism is related to energy losses into adjacent media, like from a pipe into the contained fluid (Fig. 2.4a) or structural attenuation at stiffeners, ribs and joints (Fig. 2.4b) and even into air. This mechanism is commonly known as "leakage". This factor cannot be neglected for real engineering structures, which are usually quite complex in comparison to the simple plates studied in laboratory conditions and mostly reported in the literature.

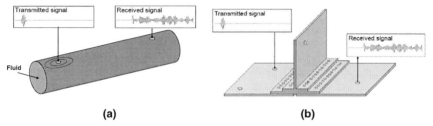

(a)   (b)

Fig. 2.4: Wave amplitude losses into adjacent media:

(a) losses in pipes due to the contained fluid; (b) structural attenuation at stiffeners and joints

The last attenuation mechanism, the attenuation due to dispersion, occurs due to different velocities for different frequency components. Therefore an initially short, broad band, pulse, begins to spread in time as the distance of propagation increases, as shown in Fig. 2.5. This causes a loss in amplitude. The magnitude of amplitude loss depends on the steepness of the dispersion curves and the bandwidth of the signal. This type of attenuation has been little studied in the literature, since mostly narrow band excitation in frequency ranges with little dispersion is conducted.

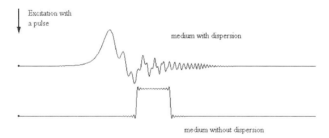

Fig. 2.5: Amplitude losses due to velocity dispersion [Kro99]

As already mentioned, the focus of this thesis is to study the attenuation due to material damping.

2.3 Fundamentals of piezoelectric materials and PWAS

Since this work uses piezoelectric active sensors (PWAS) both as strain sensors as well as strain exciters, a short review on piezoelectric materials is given next.

The piezoelectric effect was discovered in 1880 by the brothers Jacques and Pierre Curie. The word *piezo* is a Greek word which means to squeeze or to press. Piezo-electricity is the ability of some materials to generate an electric charge due to applied mechanical stress. This effect is reversible, i.e. the piezoelectric material responds with a mechanical strain to an applied electric field. In this work, thin piezo-electric wafers (Fig. 2.6) are used.

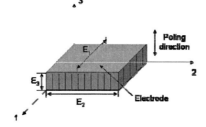

Fig. 2.6: Piezoelectric material wafer

There are a number of different materials with piezoelectric properties, the two most widely used materials being the ceramic Lead Zirconate Titanate (PZT) and the polymer Polyvinylidene Fluoride (PVDF). In their direct effect, piezoelectric materials can be used as sensors to measure structural deformations, whereas in their converse effect the materials can be used as actuators, for example for the excitation of Lamb waves for structural health monitoring applications. PZT has a Young's modulus comparable to that of aluminum, whereas the modulus of PVDF is approximately $1/12^{th}$ that of aluminum. Because of the lower stiffness, PVDF is much more suited to sensing application since it is less likely to influence the dynamics of the host structure. However due to its much higher piezoelectric coefficients, PZT is much more suitable for actuating applications. Tab. 2.1 shows typical properties of a PVDF and a PZT material (poling direction is the 3rd direction).

Material	PZT (APC-850) [Apc07]	PVDF [Sir00]
Young's Modulus		
$Y_{11}^E \, [N/mm^2]$	63000	4000 – 6000
$Y_{33}^E \, [N/mm^2]$	54000	4000 – 6000
Piezoelectric Voltage Constant		
d_{31} [m/V]	$-175 \cdot 10^{-12}$	$(18-24) \cdot 10^{-12}$
d_{32} [m/V]	$-175 \cdot 10^{-12}$	$(2.5-3) \cdot 10^{-12}$
d_{33} [m/V]	$400 \cdot 10^{-12}$	$-33 \cdot 10^{-12}$
Dielectric permittivity		
$\varepsilon_{33}^\sigma \, [F/m]$	$15.47 \cdot 10^{-9}$	$0.106 \cdot 10^{-9}$

Tab. 2.1: Comparison of typical properties of PZT and PVDF

According to [lee87] the constitutive relations for a piezoelectric material are:

$$\varepsilon_{ij} = s_{ijkl}^E \cdot \sigma_{kl} + d_{kij} \cdot E_k \qquad \text{("actuator equation")} \qquad (2.5)$$

$$D_j = d_{jkl} \cdot \sigma_{kl} + \varepsilon_{jk}^\sigma \cdot E_k \qquad \text{("sensor equation")} \qquad (2.6)$$

where ε_{ij} is the mechanical strain tensor, s_{ijkl}^E is the mechanical compliance tensor measured at zero electric field, σ_{kl} is the mechanical stress tensor, d_{kij} denotes the piezoelectric coupling effects, E_k is the electric field, D_j represents the electric displacement, and ε_{jk}^σ is the dielectric permittivity at zero stress.

<u>Conversion of mechanical strain to voltage output for sensing applications</u>

In this work thin and small piezoelectric wafer active sensors (PWAS) of PZT material are used, which utilize the d_{31} coupling between in-plane strain and transverse electric field. To use these PWAS for strain sensing application, the conversion of mechanical strain to voltage output to has to be derived. Therefore the charge q and the voltage generated across the sensor electrodes V_c are related by the capacitance of the sensor, C_p as

$$V_c = \frac{q}{C_p} \qquad (2.7)$$

According to [Sir00] a piezoelectric sheet can be treated as a parallel plate capacitor, with a capacitance given by:

$$C_p = \frac{\varepsilon_{33}^\sigma \cdot l_c \cdot b_c}{t_c} \qquad (2.8)$$

where the coefficients l_c, b_c and t_c are length, width, and thickness of the piezoelectric sensor and ε_{33}^σ denotes the dielectric permittivity. The collected q charge is related to the electric displacement and can be expressed as:

$$q = \iint [D_1 \quad D_2 \quad D_3] \begin{bmatrix} dA_1 \\ dA_2 \\ dA_3 \end{bmatrix} \qquad (2.9)$$

For the case in which there is no external electric field applied (sensor application) the electric displacement can be related to the applied mechanical stress as:

$$
\begin{bmatrix} D_1 \\ D_2 \\ D_3 \end{bmatrix} = \begin{bmatrix} 0 & 0 & 0 & 0 & d_{15} & 0 \\ 0 & 0 & 0 & d_{24} & 0 & 0 \\ d_{31} & d_{32} & d_{33} & 0 & 0 & 0 \end{bmatrix} \cdot \begin{bmatrix} \sigma_1 \\ \sigma_2 \\ \sigma_3 \\ \sigma_4 \\ \sigma_5 \\ \sigma_6 \end{bmatrix}
\tag{2.10}
$$

In the case of a PZT material, which is polarized in thickness direction (where $d_{31} = d_{32}$ and $d_{15} = d_{24}$), according to Fig. 2.6, the equations for the conversion of strain to voltage output can be derived, for the case of surface bonded sensors to the following relations:

$$
V_0 = \frac{q}{C_p} = \frac{t_c}{e_{33}^\sigma \cdot l_c \cdot b_c} \cdot \iint D_3 \, dx \, dy = \frac{d_{31} \cdot t_c}{e_{33}^\sigma \cdot l_c \cdot b_c} \cdot \iint (\sigma_{11} + \sigma_{22}) \, dx \, dy =
$$

$$
= \frac{d_{31} \cdot t_c \cdot Y_c}{e_{33}^\sigma \cdot l_c \cdot b_c \cdot (1-\nu)} \cdot \iint (\varepsilon_{11} + \varepsilon_{22}) \, dx \, dy
\tag{2.11}
$$

Assuming the strain to be averaged over the effective length and width of the piezo-ceramic sensor, the equation relating strain and voltage generated by the sensor is:

$$
V_C = \frac{t_c \cdot d_{31} \cdot Y_c}{e_{33}^\sigma \cdot (1-\nu)_c} \cdot \frac{l_{c_{eff}}}{l_c} \cdot \frac{b_{c_{eff}}}{b_c} \cdot (\varepsilon_{11} + \varepsilon_{22})
\tag{2.12}
$$

The equation (2.12) shows that for surface bonded piezoelectric sensors the voltage output V_C is proportional to the sum of the in plane normal strains ε_{11} and ε_{22}. Hence it is not possible to separate out each strain component from the voltage output. Due to shear lag effects caused by the finite thickness bond layer between PWAS and the host structure the effective dimensions of the PWAS ($l_{c_{eff}}$ and $b_{c_{eff}}$) are smaller than the geometric dimensions l_c and b_c. Assuming a structure loaded in pure bending, the analytical equation that describes the ratio of PWAS strain to the host structure strain is given as follows [Sir00]:

$$
\frac{\varepsilon_c}{\varepsilon_b} = \frac{\cosh(\Gamma \cdot l_c) - 1}{\sinh(\Gamma \cdot l_c)} \cdot \sinh(\Gamma \cdot x) - \cosh(\Gamma \cdot x) + 1
\tag{2.13}
$$

where $\Gamma = \sqrt{\dfrac{G}{Y_{11}^E \cdot t_c \cdot t_s} + \dfrac{3 \cdot b_c \cdot G}{Y_b \cdot b_b \cdot t_b \cdot t_s}}$, with G as shear modulus of the bond layer material, Y_b, b_b, t_b the Young's modulus, width and thickness of the host structure, t_s, the thickness of the bond layer, t_c, l_c and b_c the thickness, the length, and the width of the PWAS. To quantify the effect of shear lag, the equation (2.13) is plotted for an adhesive layer with $G_b = 2\ GPa$, a layer thickness of $t_s = 0.03\ mm$ and a PWAS thickness of $t_c = 0.2\ mm$. The host structure considered for the calculation of the values illustrated in Fig. 2.7 corresponds to an unidirectional carbon fiber reinforced composite, where 0° denotes the fiber direction and 90° the direction perpendicular to the fibers.

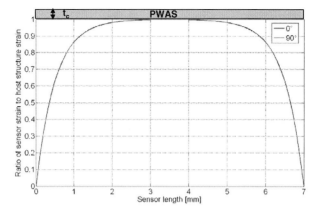

Fig. 2.7: Shear lag effect

By this approach it is now possible to define new, effective PWAS dimensions $l_{C \, eff}$ and $b_{C \, eff}$ by integrating the area under the curve in Fig. 2.7. For the values under discussion the effective length fraction (=effective width fraction) is equal to $\frac{l_{c_{eff}}}{l_c} = \frac{b_{c_{eff}}}{b_c} = 0.83$, that means 83% of the sensor length and of the sensor width contributes fully to the measured strain.

2.4 Signal processing

The intention of this section is to provide the theoretical and mathematical background for different signal processing methods which are used in this work and are crucial elements for the understanding and interpretation of signal data. Signal processing, of course, does not provide any different data or information, it only helps to distinguish useful data from not useful data, thus extracting the useful information from the raw data, i.e. in the case of this work, the signals collected by PWAS. The objectives of signal processing are signal denoising, time-frequency analysis, time-of-flight (TOF) extraction, and reliable measurements of wave amplitudes.

2.4.1 Signal denoising

Data collected from sensors are in general corrupted by noise. Since noise is unavoidable it is necessary to include a denoising procedure in the pre-processing to improve and in many cases even to enable a successful signal processing. Many different signal denoising techniques using traditional digital filters and discrete wavelet transform (DWT) have been proposed and applied in the literature. However, it is not the intention neither of this chapter nor of this research to study and give a comprehensive review on different denoising techniques. Therefore the technique which gave the best results in this research, namely the DWT denoising technique, is considered. In this research the DWT denoising technique has been chosen, since it seemed to be the most efficient procedure.

Fig. 2.8 shows a signal gathered at a PWAS due to an impact event. The signal was cleaned by DWT denoising using Meyer wavelet at level 7.

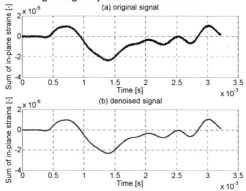

Fig. 2.8: DWT denoising of impact induced strain data gathered by PWAS:

(a) original data; (b) denoised data

2.4.2 Short Time Fourier Transform (STFT)

Since the conventional Fast Fourier Transform (FFT) assumed the signal to be stationary or it reduces the signal to stationarity, the FFT of transient, non-stationary signals is not appropriate. That is, it does not provide enough information of the content of frequencies over time. Therefore, an approach is necessary which enables the representation of transient signals in the time-frequency domain.

In the past decades, many different methods and approaches have been published, most of them being reviewed in [Coh89]. One of the methods proposed by Gabor in 1946 [Gab46] consists in windowing the signal in the time-domain and providing an FFT for each time window. Gabor's approach is known as Short Time Fourier Transform (STFT), sometimes also called windowed Fourier transform. Fig. 2.9 illustrates the procedure of windowing and mapping a signal from the time-domain into the time-frequency domain.

Fig. 2.9: Windowing of signals for short-time Fourier transform [MATLAB 2005]

Mathematically, the STFT function is defined as:

$$S_{STFT}(\omega,t) = \frac{1}{2 \cdot \pi} \int_{-\infty}^{\infty} e^{-i \cdot \omega \cdot \tau} \cdot x(\tau) \cdot h(\tau - t) d\tau \tag{2.14}$$

where $h(t)$ is a window function. The use of different window types is theoretically possible such as rectangular, Hanning or Kaiser. In this research a Hanning window is used for STFT. The disadvantage of the STFT approach is that the time vs. frequency resolution depends on the size of the window. Therefore, as illustrated in Fig. 2.9, by using a narrow window, good time resolution can be obtained (but worse frequency resolution), whereas a wide window worsens the time resolution by improves the frequency resolution. However the best compromise between time and frequency resolution has to be chosen for each application and each case.

2.4.3 Wavelet Transform (WT)

The advantage of the Wavelet Transform (WT) over STFT is the use of wavelet functions instead of a fixed window function. The WT provides a wide window in the low frequency range and a narrow window at high frequencies. This gives rise to a good frequency resolution at low frequencies and good time resolution at high frequencies (Fig. 2.10).

Fig. 2.10: Windowing of signals for Wavelet transform [MATLAB 2005]

The continuous wavelet transform (CWT) of a signal $x(t)$ is defined according to Chui [Chu92] as:

$$S_{CWT}(a,b) = \frac{1}{\sqrt{a}} \int_{-\infty}^{\infty} x(\tau) \cdot \Psi\left(\frac{\tau - b}{a}\right) d\tau \tag{2.15}$$

where a is the scale and b the position variable. The function Ψ is the analyzing wavelet, often also called mother wavelet. In this study, the Gabor function is adopted since according to [Chu92] the use of this function as mother wavelet provides a very good compromise between time and frequency resolution. The Gabor function can be expressed as [Kis95]:

$$\Psi_g(t) = \frac{1}{\sqrt[4]{\pi}} \cdot \sqrt{\frac{\omega_0}{\gamma}} \cdot \exp\left[-\frac{1}{2} \cdot \left(\frac{\omega_0 \cdot t}{\gamma}\right)^2 + i \cdot \omega_0 \cdot t\right] \tag{2.16}$$

According to [Kis95], the positive constants γ and ω_0 are set to $\gamma = \pi \cdot \sqrt{2/\ln 2}$ and $\omega_0 = 2 \cdot \pi$.

2.4.4 Hilbert Transform (HT)

The Hilbert Transform (HT) is a useful method in extracting the envelope of a signal, such as a Lamb wave packet as used in this research. It is important to extract the envelope of the guided waves in order to determine their arrival times and therefore the time-of-flight (TOF). The detection of TOF is crucial for the detection of wave velocities and wave reflections and to determine their amplitudes. Fig. 2.11 shows the envelope obtained by the Hilbert transform of a tone burst signal, used later in the experimental tests.

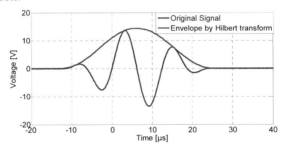

Fig. 2.11: Hilbert transform for envelope extraction

Mathematically the Hilbert transform is defined as:

$$H(x(t)) = \frac{1}{\pi} \int_{-\infty}^{\infty} x(\tau) \cdot \frac{1}{t-\tau} \, d\tau \qquad (2.17)$$

An analytical signal can be built by using the original signal $x(t)$ as the real part and the Hilbert transform $H(x(t))$ as imaginary part. This equation can be expressed as:

$$x_A(t) = x(t) + i \cdot H(x(t)) \qquad (2.18)$$

The envelope can be obtained as:

$$e(t) = \sqrt{x(t)^2 + H(x(t))^2} \qquad (2.19)$$

Thus, the envelope of Lamb wave signals has been obtained in this work by using the Hilbert transform and the processing has been provided by the mathematical software MATLAB®.

3 Development of a theoretical model for Lamb wave propagation in viscoelastic composites

For the investigation of Lamb wave propagation in layered media, such as composites, different methods can be adopted. The different approaches are (a) "exact" three dimensional (3D) solutions, (b) approximate solutions obtained by laminated plate theories of different orders and (c) semi-analytical finite element (SAFE) methods.

The exact 3D solutions are based on the superposition of bulk waves including matrix-based methods. The most common matrix methods are the Transfer Matrix method and the Global Matrix method. The basic principle of the Transfer Matrix method, which was first described by Thomson [Tho50], consists in condensing the multilayered structure into a set of four equations relating the boundary conditions at the first interface to the boundary conditions at the last interface. More details on this method can be found in the review paper by Lowe [Low95]. The Global Matrix method, which was first published by Knopoff [Kno64], assembles directly a single matrix representing the whole structure, where the system matrix consists of $4(n-1)$ equations. This method has been implemented by Pavlakovic et al. [Pav97] into a software called "DISPERSE" for obtaining dispersion curves for composites. Of course, both matrix methods have their advantages and disadvantages, which are also discussed in [Low95]. Furthermore, 3D exact solutions for elastic wave propagation have been derived by Nayfeh and Chimenti, reported in [Nay89], and extended in [Nay91] for the application in composites.

A different group of methods, a semi-analytical method, for modeling Lamb waves is the SAFE (Semi-analytical finite element) method, which was demonstrated for the first time by Lagasse [Lag73] and by Aalami [Aal73]. Such methods are also called spectral or waveguide finite element methods. The SAFE approach, in general, uses a finite element discretization of the cross-section of the waveguide for extracting dispersive Lamb wave solutions. The displacements along the wave propagation direction are described in an analytical way as harmonic functions. The dispersion curves are then typically obtained by solving an eigenvalue problem. Recently, the SAFE method was extended by Bartoli et al. [Bar06] for modeling dispersive solutions in waveguides of arbitrary cross-sections by considering wave attenuation due to material damping.

A third group of methods for modeling Lamb waves is by plate theories; this approach is used in this research. However, in order to describe the motions of Lamb waves, classical plate theories are not accurate enough; therefore higher order plate theories have to be adopted. Plate theories up to third-order have been proposed for elastic wave propagation by several researchers [Red99]. However, to the knowledge of the author of this thesis, there are no implementations of material damping in the plate theories included in the literature.

The present research extends a higher order plate theory to the modeling of Lamb waves in viscoelastic fiber-reinforced polymer composites. Comparing to "exact" 3D methods, higher order plate theories offer high computational savings and much more stable solutions, while maintaining the accuracy of the solution in the lower frequency range. Dispersion and attenuation calculations using higher order plate theories are almost real-time computation, since just polynomial equations rather than the transcendental equations of 3-D elasticity have to be solved. This method is developed in this research for approximating lower order guided wave modes (S_0, SH_0 and A_0) including anisotropic material damping.

3.1 Material damping model

As already discussed in section 2.2, there are several factors and influences on the attenuation of Lamb waves. We focus on the attenuation in Lamb waves due to internal damping or material damping in viscoelastic, anisotropic layered materials. This attenuation of waves is particularly relevant to the SHM of high-loss materials such as composites, since it decides how far Lamb waves are able to propagate. Attenuation models are moreover needed for the optimization of sensor numbers and their location.

There are two models which can be used to describe the viscoelastic, damping behaviour of materials. The first one, the hysteretic model, reviewed by Lakes [Lak99], considers complex components of the stiffness matrix of the material:

$$C_{ij}^{*} = C_{ij}^{'} + i \cdot C_{ij}^{''} \tag{3.1}$$

where $C_{ij}^{'}$ and $C_{ij}^{''}$ are the real and the imaginary part of the viscoelastic, complex stiffness tensor.

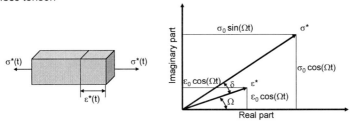

Fig. 3.1: Physical interpretation of the complex stiffness method

The hysteretic model assumes no frequency dependence of the viscoelastic constants. Another model is the Kelvin-Voigt model, which assumes a linear dependence of $C_{ij}^{''}$ coefficients with frequency and can be expressed as:

$$C_{ij}^{''} = \omega \cdot \eta_{ij} \tag{3.2}$$

where η_{ij} denotes the viscosity tensor. The use of the Kelvin-Voigt model would require a much higher calculation effort and would bring much more difficulties on the implementation in a mathematical model. For the Kelvin-Voigt model the imaginary parts of the stiffness coefficients have to be recalculated at each frequency. This increases the calculation costs considerably. Moreover, one has also to keep in mind that the material properties required for any damping model need a reliable non destructive characterization technique. However, the hysteretic model has proved to be more frequently used in composites NDE literature [Lak99], because it has the advantage that in the wave propagation model the only modification which has to be done in comparison to non-absorbing, elastic materials is that the stiffness tensor becomes complex. Therefore, in this research work, the hysteretic model is adopted.

3.2 Classical plate theory

In order to predict the dispersion and attenuation properties of guided waves in anisotropic composites, a mechanical/mathematical wave propagation model is needed. Since the focus of this research work is on thin, plate-like structures, plate theories can be used. The most simple plate theory is the classical plate theory (CPT), which can be extended to include the anisotropic viscoelastic properties predicted by laminated plate theory. The dispersion and attenuation relations for the axial and flexural waves can be obtained by using the equations of motions. However, it must be pointed out particularly, that at higher wave frequencies, the use of CPT is not appropriate for the prediction of dispersion and attenuation relations of Lamb waves. In those cases, more accurate results can be obtained by a higher order plate theory (HPT). Fig. 3.2 illustrates the dispersion behavior for a 1-mm thick aluminum plate obtained by the CPT and HPT. Thus, the axial and flexural waves predicted by the CPT are only low-frequency approximations of the Lamb wave S_0 and A_0 modes. This can be traced back to the fact that the CPT neglects the effects of transverse shear deformation and rotary inertia which become important at higher frequencies.

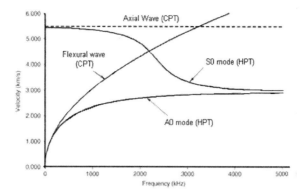

Fig. 3.2: Classical plate theory (CPT) vs. Higher order plate theory HPT [Pro91]

For axial mode waves in a symmetric orthotropic laminate, the predictions of the velocity are derived from the equations of motion for in-plane displacements which are given by [Pro91]

$$A_{11}^*\left(\frac{\partial^2 u_0}{\partial x^2}\right) + A_{66}^*\left(\frac{\partial^2 u_0}{\partial y^2}\right) + \left(A_{12}^* + A_{66}^*\right)\left(\frac{\partial^2 v_0}{\partial x \partial y}\right) = \rho h\left(\frac{\partial^2 u_0}{\partial t^2}\right) \tag{3.3}$$

and

$$A_{22}^* \left(\frac{\partial^2 v_0}{\partial y^2} \right) + A_{66}^* \left(\frac{\partial^2 v_0}{\partial x^2} \right) + \left(A_{12}^* + A_{66}^* \right) \left(\frac{\partial^2 u_0}{\partial x \partial y} \right) = \rho h \left(\frac{\partial^2 v_0}{\partial t^2} \right) \qquad (3.4)$$

In the case of viscoelastic material behavior adopting the hysteretic model presented in Chapter 3.1, the in-plane stiffness coefficients (A_{ij}^*) obtained by the lamination theory [Ash70] become complex. The above equations predict two modes of propagation. In general, one mode is quasi-axial with the largest component of its particle displacement along the propagation direction and the other is quasi-in-plane shear with the largest component of particle displacement perpendicular to the direction of propagation in the plane of the plate. Along symmetry directions, these modes are pure axial mode and in-plane shear mode. For propagation along the x axis or 0° direction, this approach assumes a pure axial mode with velocity (c_1) given

$$c_1 = \sqrt{\frac{A_{11}'}{\rho h}} \qquad (3.5)$$

and a pure in-plane shear mode with velocity (c_2) given by

$$c_2 = \sqrt{\frac{A_{66}'}{\rho h}} \qquad (3.6)$$

where h denotes the thickness of the laminate and ρ its density. For the calculation of wave velocities, only the real part A_{ij}' of the complex in-plane stiffness coefficients A_{ij} needs to be considered. For wave propagation along the y- axis or 90° direction the velocity of the pure axial mode is given by

$$c_3 = \sqrt{\frac{A_{22}'}{\rho h}} \qquad (3.7)$$

since it is also a symmetry axis of the orthotropic laminate. For the calculation of the axial wave velocity for off-axis propagation all stiffness coefficients of the laminate are involved. The following equation gives the velocity for a quasi-axial mode for 45° propagation

$$c_4 = \sqrt{\frac{(A_{11}' + 2A_{66}' + A_{22}') + \sqrt{(A_{11}' + 2A_{66}' + A_{22}')^2 - 4(A_{11}' + A_{66}')(A_{22}' + A_{66}') + 4(A_{12}' + A_{66}')^2}}{4\rho h}} \qquad (3.8)$$

As shown in the above equation and in Fig. 3.2, the axial waves predicted by the CPT are not dispersive, that is, the wave speed is not a function of frequency.

For flexural waves, CPT can be also used to predict the dispersion relations. In this case, the plate is assumed to be under a state of pure bending in which plane sections of the plate remain plane and perpendicular to the mid-plane of the plate (Kirchhoff theory). Thus, shear deformation is not included in this theory. A state of plane stress is assumed and the effects of rotary inertia are also neglected. The CPT equation of motion for an orthotropic composite laminate in the absence of body forces can be expressed as

$$D_{11}^*\left(\frac{\partial^4 w}{\partial x^4}\right)+4D_{16}^*\left(\frac{\partial^4 w}{\partial x^3 \partial y}\right)+2\left(D_{12}^*+2D_{66}^*\right)\left(\frac{\partial^4 w}{\partial x^2 \partial y^2}\right)+4D_{26}^*\left(\frac{\partial^4 w}{\partial x \partial y^3}\right)+D_{22}^*\left(\frac{\partial^4 w}{\partial y^4}\right)+\rho h\left(\frac{\partial^2 w}{\partial t^2}\right)=0$$

(3.9)

Where the D_{ij}' values are the real parts of the complex bending stiffness coefficients obtained from the lamination theory [Ash70]. In the above equation, w denotes the (out-of-plane) displacement along the z axis or normal to the plane of the plate. The dispersion behavior for the flexural mode predicted using CPT is obtained by substituting the displacement for a plane wave propagating in an arbitrary direction into the equation of motion. This assumption for the displacement is of the form

$$w_0 = W_0 \cdot e^{i \cdot \left[k^* \cdot (l_1 \cdot x + l_2 \cdot y) - \omega t\right]}$$

(3.10)

Thus for the case of viscoelastic material, where the waves are attenuated, the wave number $k^* = k' - i k''$ has also to become complex. The angular frequency ω is a real positive value. Fig. 3.3 shows the influence of the real and the imaginary part of the wave number. Thus the imaginary part of the complex wave number denotes the wave attenuation factor with units 1/(distance unit).

$$w_0 = W_0 \cdot e^{i \cdot \left[k^* \cdot (l_1 \cdot x + l_2 \cdot y) - \omega \cdot t\right]} = W_0 \cdot e^{i \cdot \left[k' \cdot (l_1 \cdot x + l_2 \cdot y) - \omega \cdot t\right]} \cdot e^{-k'' \cdot (l_1 \cdot x + l_2 \cdot y)}$$

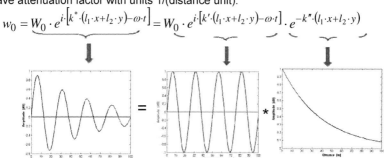

Fig. 3.3: Components of displacement field assumptions

After substitution and reduction of terms, the resulting CPT dispersion relation is found to be [Pro91]:

$$c_f = \sqrt[4]{\frac{D_{11}' \cdot l_1^4 + 4 \cdot D_{16}' \cdot l_1^3 \cdot l_2 + 2 \cdot \left(D_{12}' + 2 \cdot D_{66}'\right) \cdot l_1^2 \cdot l_2^2 + 4 \cdot D_{26}' \cdot l_1 \cdot l_2^3 + D_{22}' \cdot l_2^4}{\rho \cdot h}} \cdot \sqrt{\omega}$$

(3.11)

In summary, CPT contains two non-dispersive symmetric modes (axial and in-plane shear mode) and one dispersive anti-symmetrical (flexural) mode. The axial and flexural wave modes obtained by CPT are low-frequency approximations of the low order Lamb wave modes S_0 and A_0 (Fig. 3.2). Reference [Lih95] mentions that CPT is generally recognized to be accurate enough only if the wavelength is larger than about ten times the laminate thickness, that is for the CFRP laminates analyzed in this work up to a frequency of about 20 kHz. Since in this research Lamb waves with a frequency up to 400 kHz are investigated, a higher order plate theory (HPT) is needed, which is highlighted in the next section.

3.3 Higher order plate theory

The so-called Mindlin plate theory [Min51] is based on the CPT but takes additionally shear and rotatory inertia into account. The Mindlin plate theory can be also referred to as first-order shear deformation plate theory (SDPT). Compared to Mindlin plate theory (or SDPT) a higher-order plate theory (HPT) is able to represent the kinematics better, and in the case of Lamb waves, particularly the mode shapes. Of course, compared to Mindlin's theory, the HPT requires much more computational effort, however it represents a good compromise between accuracy and computational effort, thus between the Mindlin plate theory and the exact three-dimensional (3D) solution. This is the motivation in this research for adopting a higher-order plate theory for the theoretical prediction of dispersion and attenuation properties of Lamb waves in viscoelastic composites.

The theory of reference in this study is reported in [Whi73] and is extended in this work by appending up to second-order terms to the displacement field assumption.

3.3.1 Constitutive relations

An infinite plate composed of anisotropic thin plies perfectly bonded together, according to the assumptions of the lamination theory [Ash70] is considered, where the z-direction denotes the out-of-plane direction being normal to the laminate's surface. Fig. 3.4 shows the definition of forces and moment resultants, as will be used for the derivation of the analytical model for wave propagation.

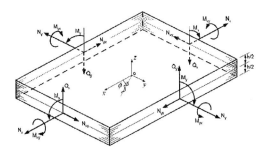

Fig. 3.4: Forces and moments of a composite plate [Wan04]

The Hooke's law for a monoclinic (with respect to z-axis), viscoelastic material such of a unidirectional ply with an arbitrary orientation can be expressed as:

$$
\begin{Bmatrix}
\sigma_x^* \\
\sigma_y^* \\
\sigma_z^* \\
\tau_{yz}^* \\
\tau_{xz}^* \\
\tau_{xy}^*
\end{Bmatrix} =
\begin{bmatrix}
C_{11}^* & C_{12}^* & C_{13}^* & 0 & 0 & C_{16}^* \\
C_{12}^* & C_{22}^* & C_{23}^* & 0 & 0 & C_{26}^* \\
C_{13}^* & C_{23}^* & C_{33}^* & 0 & 0 & C_{36}^* \\
0 & 0 & 0 & C_{44}^* & C_{45}^* & 0 \\
0 & 0 & 0 & C_{45}^* & C_{55}^* & 0 \\
C_{16}^* & C_{26}^* & C_{36}^* & 0 & 0 & C_{66}^*
\end{bmatrix} \cdot
\begin{Bmatrix}
\varepsilon_x^* \\
\varepsilon_y^* \\
\varepsilon_z^* \\
\gamma_{yz}^* \\
\gamma_{xz}^* \\
\gamma_{xy}^*
\end{Bmatrix}
\tag{3.12}
$$

where c_{ij}^* denotes the complex stiffness coefficients of a monoclinic lamina. In the case of a viscoelastic material both the stresses and the strains are also complex. According to the lamination theory [Ash70] stress and moments per unit length are introduced as follows

$$
(N_x^*, N_y^*, N_z^*, N_{xy}^*, Q_x^*, Q_y^*) = \int_{-h/2}^{h/2} (\sigma_x^*, \sigma_y^*, \sigma_z^*, \tau_{xy}^*, \tau_{xz}^*, \tau_{yz}^*)\, dz
\tag{3.13}
$$

$$
(M_x^*, M_y^*, M_{xy}^*, R_x^*, R_y^*) = \int_{-h/2}^{h/2} (\sigma_x^*, \sigma_y^*, \tau_{xy}^*, \tau_{xz}^*, \tau_{yz}^*) z\, dz
\tag{3.14}
$$

$$
(S_x^*, S_y^*, S_{xy}^*) = \frac{1}{2} \int_{-h/2}^{h/2} (\sigma_x^*, \sigma_y^*, \tau_{xy}^*)\, dz
\tag{3.15}
$$

As reported in [Cas01], the mode shapes of guided waves (especially at higher frequencies) cannot be assumed to have a linear strain distribution across the thickness. Therefore it is necessary for the accuracy of the wave solution to extend the strain and displacement field to higher order terms. Thus, the displacement field is approximated in this work with terms up to second order for the in-plane displacements u, v and for the out-of-plane displacement w up to first order terms obtaining the following form

$$
u^*(x, y, z, t) = u_0^*(x, y, t) + z \psi_x^*(x, y, t) + \frac{z^2}{2} \phi_x^*(x, y, t)
\tag{3.16}
$$

$$
v^*(x, y, z, t) = v_0^*(x, y, t) + z \psi_y^*(x, y, t) + \frac{z^2}{2} \phi_y^*(x, y, t)
\tag{3.17}
$$

$$
w^*(x, y, z, t) = w_0^*(x, y, t) + z \psi_z^*(x, y, t)
\tag{3.18}
$$

where u_0, v_0 and w_0 represent the displacements of every point of the mid-plane of the plate, ψ_x and ψ_y are the rotations of section x = constant and y = constant, respectively. By assuming the displacement field above, the following relations for the strain field are obtained

$$\varepsilon_x^* = \frac{\partial u_0^*}{\partial x} + z\left(\frac{\partial \psi_x^*}{\partial x}\right) + \frac{z^2}{2}\left(\frac{\partial \phi_x^*}{\partial x}\right) \tag{3.19}$$

$$\varepsilon_y^* = \frac{\partial v_0^*}{\partial y} + z\left(\frac{\partial \psi_y^*}{\partial y}\right) + \frac{z^2}{2}\left(\frac{\partial \phi_y^*}{\partial y}\right) \tag{3.20}$$

$$\varepsilon_z^* = \psi_z^* \tag{3.21}$$

$$\gamma_{xy}^* = \frac{\partial u_0^*}{\partial y} + \frac{\partial v_0^*}{\partial x} + z\left(\frac{\partial \psi_x^*}{\partial y} + \frac{\partial \psi_y^*}{\partial x}\right) + \frac{z^2}{2}\left(\frac{\partial \phi_x^*}{\partial y} + \frac{\partial \phi_y^*}{\partial x}\right) \tag{3.22}$$

$$\gamma_{xz}^* = \psi_x^* + \frac{\partial w_0^*}{\partial x} + z\left(\phi_x^* + \frac{\partial \psi_z^*}{\partial x}\right) \tag{3.23}$$

$$\gamma_{yz}^* = \psi_y^* + \frac{\partial w_0^*}{\partial y} + z\left(\phi_y^* + \frac{\partial \psi_z^*}{\partial y}\right) \tag{3.24}$$

Since, in plate theories, the transverse shear stiffness is overestimated compared to the exact 3D solution and to the real stiffness, shear correction factors κ_i are introduced in this research in a similar manner to that adopted by Reissner [Rei45] and Mindlin [Min59]. Since the accurate consideration of transverse shear stiffness in plate theories is a challenging and crucial problem, in particular in combination with composite materials, section 3.3.4 is dedicated to shear correction factors. The transversal shear strains $\gamma_{xz}^*, \gamma_{yz}^*$ and the out-of-plane normal strain ε_z^* have therefore to be reduced by shear correction factors. These strains can be then written in the following form

$$\varepsilon_z^* = \kappa_3 \cdot \psi_z^* \tag{3.25}$$

$$\gamma_{xz}^* = \kappa_1 \cdot \left(\psi_x^* + \frac{\partial w_0^*}{\partial x}\right) + \kappa_4 \cdot z\left(\phi_x^* + \frac{\partial \psi_z^*}{\partial x}\right) \tag{3.26}$$

$$\gamma_{yz}^* = \kappa_2 \cdot \left(\psi_y^* + \frac{\partial w_0^*}{\partial y}\right) + \kappa_5 \cdot z\left(\phi_y^* + \frac{\partial \psi_z^*}{\partial y}\right) \tag{3.27}$$

Substituting equation (3.19 – 3.24) into (3.12) and introducing their results into equations (3.13 – 3.15) the following constitutive relation is obtained:

$$
\begin{bmatrix} N_x^* \\ N_y^* \\ N_z^* \\ N_{xy}^* \\ M_x^* \\ M_y^* \\ M_{xy}^* \\ S_x^* \\ S_y^* \\ S_{xy}^* \end{bmatrix} =
\begin{bmatrix}
A_{11}^* & A_{12}^* & \kappa_3 A_{13}^* & A_{16}^* & B_{11}^* & B_{12}^* & B_{16}^* & D_{11}^*/2 & D_{12}^*/2 & D_{16}^*/2 \\
A_{12}^* & A_{22}^* & \kappa_3 A_{23}^* & A_{26}^* & B_{12}^* & B_{22}^* & B_{26}^* & D_{12}^*/2 & D_{22}^*/2 & D_{26}^*/2 \\
\kappa_3 A_{13}^* & \kappa_3 A_{23}^* & \kappa_3^2 A_{33}^* & \kappa_3 A_{36}^* & \kappa_3 B_{13}^* & \kappa_3 B_{23}^* & \kappa_3 B_{36}^* & \kappa_3 D_{13}^*/2 & \kappa_3 D_{23}^*/2 & \kappa_3 D_{36}^*/2 \\
A_{16}^* & A_{26}^* & \kappa_3 A_{36}^* & A_{66}^* & B_{16}^* & B_{26}^* & B_{66}^* & D_{16}^*/2 & D_{26}^*/2 & D_{66}^*/2 \\
B_{11}^* & B_{12}^* & \kappa_3 B_{13}^* & B_{16}^* & D_{11}^* & D_{12}^* & D_{16}^* & F_{11}^*/2 & F_{12}^*/2 & F_{16}^*/2 \\
B_{12}^* & B_{22}^* & \kappa_3 B_{23}^* & B_{26}^* & D_{12}^* & D_{22}^* & D_{26}^* & F_{12}^*/2 & F_{22}^*/2 & F_{26}^*/2 \\
B_{16}^* & B_{26}^* & \kappa_3 B_{36}^* & B_{66}^* & D_{16}^* & D_{26}^* & D_{66}^* & F_{16}^*/2 & F_{26}^*/2 & F_{66}^*/2 \\
D_{11}^*/2 & D_{12}^*/2 & \kappa_3 D_{13}^*/2 & D_{16}^*/2 & F_{11}^*/2 & F_{12}^*/2 & F_{16}^*/2 & H_{11}^*/4 & H_{12}^*/4 & H_{16}^*/4 \\
D_{12}^*/2 & D_{22}^*/2 & \kappa_3 D_{23}^*/2 & D_{26}^*/2 & F_{12}^*/2 & F_{22}^*/2 & F_{26}^*/2 & H_{12}^*/4 & H_{22}^*/4 & H_{26}^*/4 \\
D_{16}^*/2 & D_{26}^*/2 & \kappa_3 D_{36}^*/2 & D_{66}^*/2 & F_{16}^*/2 & F_{26}^*/2 & F_{66}^*/2 & H_{16}^*/4 & H_{26}^*/4 & H_{66}^*/4
\end{bmatrix}
\cdot
\begin{bmatrix} u_{0,x}^* \\ v_{0,y}^* \\ \psi_z^* \\ u_{0,y}^* + v_{0,x}^* \\ \psi_{x,x}^* \\ \psi_{y,y}^* \\ \psi_{x,y}^* + \psi_{y,x}^* \\ \phi_{x,x}^* \\ \phi_{y,y}^* \\ \phi_{x,y}^* + \phi_{y,x}^* \end{bmatrix}
\tag{3.28}
$$

and

$$\begin{Bmatrix} Q_x^* \\ Q_y^* \\ R_x^* \\ R_y^* \end{Bmatrix} = \begin{bmatrix} \kappa_1^2 A_{55}^* & \kappa_1 \kappa_2 A_{45}^* & \kappa_1 \kappa_4 B_{55}^* & \kappa_1 \kappa_5 B_{45}^* \\ \kappa_1 \kappa_2 A_{45}^* & \kappa_2^2 A_{44}^* & \kappa_2 \kappa_4 B_{45}^* & \kappa_2 \kappa_5 B_{44}^* \\ \kappa_1 \kappa_4 B_{55}^* & \kappa_1 \kappa_5 B_{45}^* & \kappa_4^2 D_{55}^* & \kappa_4 \kappa_5 D_{45}^* \\ \kappa_2 \kappa_4 B_{45}^* & \kappa_2 \kappa_5 B_{44}^* & \kappa_4 \kappa_5 D_{45}^* & \kappa_5^2 D_{44}^* \end{bmatrix} \cdot \begin{Bmatrix} \psi_x^* + w_{0,x}^* \\ \psi_y^* + w_{0,y}^* \\ \psi_{z,x}^* + \phi_x^* \\ \psi_{z,y}^* + \phi_y^* \end{Bmatrix} . \tag{3.29}$$

The coefficients $A_{ij}^*, B_{ij}^*, D_{ij}^*, F_{ij}^*, H_{ij}^*$ can easily be obtained by a similar approach as in clasical lamination theory by the following equation

$$\left(A_{ij}^*, B_{ij}^*, D_{ij}^*, F_{ij}^*, H_{ij}^* \right) = \int_{-\frac{h}{2}}^{\frac{h}{2}} Q_{ij}^* \left(1, z, z^2, z^3, z^4 \right) dz \tag{3.30}$$

The terms Q_{ij}^* for $i, j = 1, 2, 6$ are the plane-stress reduced complex stiffnesses and Q_{ij}^* for $i, j = 4, 5$ are transverse shear stiffnesses of each unidirectional ply and defined as:

$$Q_{ij}^* = \left(C_{ij}^* - \left(\frac{C_{ij}^*(i,3) \cdot C_{ij}^*(3,j)}{C_{ij}^*(3,3)} \right) \right) \qquad \text{for } i, j = 1, 2, 6 \tag{3.31}$$

$$Q_{ij}^* = C_{ij}^* \qquad \text{for } i, j = 3, 4, 5 \tag{3.32}$$

The stiffness coefficients B_{ij} and F_{ij} describe the mechanical coupling between symmetric and antisymmetric modes. For symmetric laminates, these coefficients are all zero, such that the symmetric and antisymmetric wave modes can be considered as decoupled.

According to Hamilton's principle (or principle of virtual displacement) [Was82], which can be formulated as

$$\int_{t1}^{t2} (\delta U + \delta V - \delta K) dt = 0 \tag{3.33}$$

the equations of motion can be derived based on the strain-displacement relations. In equation (3.33), δU denotes the virtual strain energy, δV the virtual work done by applied force, and δK is the virtual kinetic energy.

$$\frac{\partial N_x^*}{\partial x} + \frac{\partial N_{xy}^*}{\partial y} + q_x^* = I_0 \left(\frac{\partial^2 u_0^*}{\partial t^2} \right) + \frac{I_2}{2} \left(\frac{\partial^2 \psi_x^*}{\partial t^2} \right) \tag{3.34}$$

$$\frac{\partial N_{xy}^*}{\partial x} + \frac{\partial N_y^*}{\partial y} + q_y^* = I_0 \left(\frac{\partial^2 v_0^*}{\partial t^2} \right) + \frac{I_2}{2} \left(\frac{\partial^2 \phi_y^*}{\partial t^2} \right) \tag{3.35}$$

$$\frac{\partial Q_x^*}{\partial x} + \frac{\partial Q_y^*}{\partial y} + q_z^* = I_0 \left(\frac{\partial^2 w_0^*}{\partial t^2} \right) \tag{3.36}$$

$$\frac{\partial M_x^*}{\partial x} + \frac{\partial M_{xy}^*}{\partial y} - Q_x^* + m_x^* = \frac{I_2}{2}\left(\frac{\partial^2 \psi_x^*}{\partial t^2}\right) \tag{3.37}$$

$$\frac{\partial M_{xy}^*}{\partial x} + \frac{\partial M_y^*}{\partial y} - Q_y^* + m_y^* = \frac{I_2}{2}\left(\frac{\partial^2 \psi_y^*}{\partial t^2}\right) \tag{3.38}$$

$$\frac{\partial R_x^*}{\partial x} + \frac{\partial R_y^*}{\partial y} - N_z^* + m_x^* = I_2\left(\frac{\partial^2 \psi_x^*}{\partial t^2}\right) \tag{3.39}$$

$$\frac{\partial S_x^*}{\partial x} + \frac{\partial S_{xy}^*}{\partial y} - R_x^* + n_x^* = \frac{I_0}{2}\left(\frac{\partial^2 u_0^*}{\partial t^2}\right) + \frac{I_4}{2}\left(\frac{\partial^2 \phi_x^*}{\partial t^2}\right) \tag{3.40}$$

$$\frac{\partial S_{xy}^*}{\partial x} + \frac{\partial S_y^*}{\partial y} - R_y^* + n_y^* = \frac{I_0}{2}\left(\frac{\partial^2 v_0^*}{\partial t^2}\right) + \frac{I_4}{2}\left(\frac{\partial^2 \phi_y^*}{\partial t^2}\right) \tag{3.41}$$

with $(I_{0j}, I_1, I_2, I_3, I_4) = \int\limits_{-h/2}^{h/2} \rho\,(1, z, z^2, z^3, z^4)dz$.

As mentioned before, in order to derive the displacement equations of motion an assumption for the solution form of the displacement field is necessary. By the assumption for the displacement field, already illustrated in Fig. 3.3, the following form is obtained:

$$u_0^* = U_0^*\, e^{i\left[(k_x^* x + k_y^* y) - \omega t\right]}, v_0^* = V_0^*\, e^{i\left[(k_x^* x + k_y^* y) - \omega t\right]}, w_0^* = W_0^*\, e^{i\left[(k_x^* x + k_y^* y) - \omega t\right]}, \psi_x^* = \Psi_x^*\, e^{i\left[(k_x^* x + k_y^* y) - \omega t\right]},$$

$$\psi_y^* = \Psi_y^*\, e^{i\left[(k_x^* x + k_y^* y) - \omega t\right]}, \psi_z^* = \Psi_z^*\, e^{i\left[(k_x^* x + k_y^* y) - \omega t\right]}, \phi_x^* = \Phi_x^*\, e^{i\left[(k_x^* x + k_y^* y) - \omega t\right]}, \phi_y^* = \Phi_y^*\, e^{i\left[(k_x^* x + k_y^* y) - \omega t\right]}. \tag{3.42}$$

Finally the following equation of motions are obtained:

$$\begin{bmatrix} L_{11}^* & L_{12}^* & 0 & L_{14}^* & L_{15}^* & L_{16}^* & L_{17}^* & L_{18}^* \\ L_{12}^* & L_{22}^* & 0 & L_{15}^* & L_{25}^* & L_{26}^* & L_{18}^* & L_{28}^* \\ 0 & 0 & L_{33}^* & L_{34}^* & L_{35}^* & L_{36}^* & L_{37}^* & L_{38}^* \\ L_{14}^* & L_{15}^* & L_{34}^* & L_{44}^* & L_{45}^* & L_{46}^* & L_{47}^* & L_{48}^* \\ L_{15}^* & L_{25}^* & L_{35}^* & L_{45}^* & L_{55}^* & L_{56}^* & L_{57}^* & L_{58}^* \\ L_{16}^* & L_{26}^* & L_{36}^* & L_{46}^* & L_{56}^* & L_{66}^* & L_{67}^* & L_{68}^* \\ L_{17}^* & L_{18}^* & L_{37}^* & L_{47}^* & L_{57}^* & L_{67}^* & L_{77}^* & L_{78}^* \\ L_{18}^* & L_{28}^* & L_{38}^* & L_{48}^* & L_{58}^* & L_{68}^* & L_{78}^* & L_{88}^* \end{bmatrix} \begin{bmatrix} U_0^* \\ V_0^* \\ W_0^* \\ \Psi_x^* \\ \Psi_y^* \\ \Psi_z^* \\ \Phi_x^* \\ \Phi_y^* \end{bmatrix} = 0 \tag{3.43}$$

where the coefficients L_{ij}^* are defined as:

$$L_{11}^* = A_{11}^* k_x^{*2} + 2A_{16}^* k_x^* k_y^* + A_{66}^* k_y^{*2} - \omega^2 I_0 \tag{3.44}$$

$$L_{12}^* = A_{16}^* k_x^{*2} + (A_{12}^* + A_{66}^*)k_x^* k_y^* + A_{26}^* k_y^{*2} \tag{3.45}$$

$$L_{17}^* = -i\kappa_3(A_{13}^* k_x^* + A_{36}^* k_y^*) \tag{3.46}$$

$$L_{18}^* = \frac{1}{2}(D_{11}^* k_x^{*2} + 2D_{16}^* k_x^* k_y^* + D_{66}^* k_y^{*2} - \omega^2 I_0) \tag{3.47}$$

$$L_{15}^* = \frac{1}{2}(D_{16}^* k_x^{*2} + (D_{12}^* + D_{66}^*) k_x^* k_y^* + D_{26}^* k_y^{*2}) \tag{3.48}$$

$$L_{22}^* = A_{66}^* k_x^{*2} + 2A_{26}^* k_x^* k_y^* + A_{22}^* k_y^{*2} - \omega^2 I_0 \tag{3.49}$$

$$L_{26}^* = -i\kappa_3 (A_{36}^* k_x^* + A_{23}^* k_y^*) \tag{3.50}$$

$$L_{28}^* = \frac{1}{2}(D_{66}^* k_x^{*2} + 2D_{26}^* k_x^* k_y^* + D_{22}^* k_y^{*2} - \omega^2 I_2) \tag{3.51}$$

$$L_{33}^* = -(\kappa_1^2 A_{55}^* k_x^* + 2\kappa_1 \kappa_2 A_{45}^* k_x^* k_y^* + \kappa_2^2 A_{44}^* k_y^* - \omega^2 I_0) \tag{3.52}$$

$$L_{34}^* = i\kappa_1 (\kappa_1 A_{55}^* k_x^* + \kappa_2 A_{45}^* k_y^*), \quad L_{35}^* = i\kappa_1 (\kappa_1 A_{45}^* k_x^* + \kappa_2 A_{44}^* k_y^*) \tag{3.53}$$

$$L_{44}^* = D_{11}^* k_x^{*2} + 2D_{16}^* k_x^* k_y^* + D_{66}^* k_y^{*2} + \kappa_1^2 A_{55}^* - \omega^2 I_2 \tag{3.54}$$

$$L_{45}^* = D_{16}^* k_x^{*2} + (D_{12}^* + D_{66}^*) k_x^* k_y^* + D_{26}^* k_y^{*2} + \kappa_1 \kappa_2 A_{45}^* \tag{3.55}$$

$$L_{55}^* = D_{66}^* k_x^{*2} + 2D_{26}^* k_x^* k_y^* + D_{22}^* k_y^{*2} + \kappa_2^2 A_{55}^* - \omega^2 I_2 \tag{3.56}$$

$$L_{66}^* = -\kappa_4^2 D_{55}^* k_x^{*2} - 2\kappa_4 \kappa_5 D_{45}^* k_x^* k_y^* - \kappa_5^2 D_{44}^* k_y^{*2} - \kappa_3^2 A_{33}^* + \omega^2 I_2 \tag{3.57}$$

$$L_{67}^* = -i(\kappa_4^2 D_{55}^* - \kappa_3 \frac{D_{13}^*}{2}) k_x^* + i(\kappa_4 \kappa_5 D_{45}^* - \kappa_3 \frac{D_{36}^*}{2}) k_y^* \tag{3.58}$$

$$L_{68}^* = -i(\kappa_4 \kappa_5 D_{45}^* - \kappa_3 \frac{D_{36}^*}{2}) k_x^* + i(\kappa_5^2 D_{44}^* - \kappa_3 \frac{D_{23}^*}{2}) k_y^* \tag{3.59}$$

$$L_{77}^* = \frac{1}{4}(H_{11}^* k_x^{*2} + 2H_{16}^* k_x^* k_y^* + H_{66}^* k_y^{*2} + 4\kappa_4^2 D_{55}^* - \omega^2 I_4) \tag{3.60}$$

$$L_{78}^* = \frac{1}{4}(H_{16}^* k_x^{*2} + (H_{12}^* + H_{66}^*) k_x^* k_y^* + H_{26}^* k_y^{*2} + 4\kappa_4 \kappa_5 D_{45}^*) \tag{3.61}$$

$$L_{88}^* = \frac{1}{4}(H_{66}^* k_x^{*2} + 2H_{26}^* k_x^* k_y^* + H_{22}^* k_y^{*2} + 4\kappa_5^2 D_{44}^* - \omega^2 I_4) \tag{3.62}$$

In order to obtain phase and group velocities and attenuation dispersion curves, the above equation can be formulated as an eigenvalue problem of the form

$$(L - \omega^2 I)\xi = 0 \tag{3.63}$$

where $\xi = \{U_0^* \quad V_0^* \quad W_0^* \quad \Psi_x^* \quad \Psi_y^* \quad \Psi_z^* \quad \Phi_x^* \quad \Phi_y^*\}^T$, L and I are 8 x 8 matrices, thus eight eigenvalues or eight Lamb wave modes can be obtained.

Since most of the composite laminates in engineering structures have symmetric lay-ups, only symmetrical laminates are considered in this research for simplification reasons and in order to reduce computation time. This assumption implicates that the stiffness coefficients B_{ij}^* and F_{ij}^* resulting from the lamination theory are equal to zero. Consequently, the coefficients L_{15}^*, L_{37}^*, L_{38}^*, L_{48}^*, L_{56}^*, L_{57}^* are also equal to zero, allowing therefore to decouple symmetrical and antisymmetric wave modes. For the symmetric wave modes, the following relation are obtained:

$$\begin{bmatrix} L_{11}^* & L_{12}^* & L_{16}^* & L_{17}^* & L_{18}^* \\ & L_{22}^* & L_{26}^* & L_{18}^* & L_{28}^* \\ & & L_{66}^* & L_{67}^* & L_{68}^* \\ & sym & & L_{77}^* & L_{78}^* \\ & & & & L_{88}^* \end{bmatrix} \begin{bmatrix} U_0^* \\ V_0^* \\ \Psi_z^* \\ \Phi_x^* \\ \Phi_y^* \end{bmatrix} = 0 \tag{3.64}$$

and for the antisymmetric wave modes

$$\begin{bmatrix} L_{33}^* & L_{34}^* & L_{35}^* \\ & L_{44}^* & L_{45}^* \\ sym & & L_{55}^* \end{bmatrix} \begin{Bmatrix} W_0^* \\ \Psi_x^* \\ \Psi_y^* \end{Bmatrix} = 0 \tag{3.65}$$

Thus, from these equations, eight decoupled wave modes, five symmetric (S_0, S_1, S_2, SH_0, SH_2) and three anti-symmetric (A_0, A_1, SH_1) can be obtained. However, in the framework of this research, the analysis is restricted only to the low-order wave modes A_0, SH_0 and S_0, since these wave types are most used for structural health monitoring applications and can be easily excited with PWAS.

3.3.2 Determination of Phase velocities and Attenuation factors

For a given angular frequency ω, the corresponding wave number k^* can be obtained. As already mentioned before, in order to consider material damping during the propagation of Lamb waves, a complex wave number k^* has to be introduced, in addition to complex material properties. However, the angular frequency ω remains a real value. With this assumptions the phase velocity can be obtained as

$$c_p = \frac{\omega}{|k'|} \tag{3.66}$$

The attenuation factor k'' is simply the imaginary part of the complex wave number. The value can be plotted versus frequency and by this approach attenuation dispersion curves can be obtained.

3.3.3 Determination of Group velocities

The group velocity c_g is related to the wave packet velocity or the velocity with which the variations in the shape of the wave's amplitude (known as the modulation or envelope of the wave) propagate. c_g can be expressed as the derivative of the angular frequency with respect to the wave number:

$$c_g = \frac{\partial \omega}{\partial k'} \tag{3.67}$$

For isotropic materials, the group velocity and the phase velocity have the same direction of propagation. However, in the case of anisotropic materials, the direction of the group velocity (i.e. energy flow) differs from that of the phase front direction [Nay95]. Thus, the direction of the group velocity does not coincide, in general, with the direction of the wave number vector k. With ϕ as the direction angle of the phase velocity, the following relation between ϕ and and the wavenumber k can be defined:

$$\phi = \arctan\left(\frac{k'_y}{k'_x}\right) \quad , \text{ where } \quad k' = \sqrt{k_x'^2 + k_x'^2} \tag{3.68}$$

According to the definition of group velocity, Eq. (3.67), the following relations for each direction component of the group velocity can be derived

$$C_{gx} = \frac{\partial \omega}{\partial k'_x} = \frac{\partial \omega}{\partial k'}\frac{\partial k'}{\partial k'_x} + \frac{\partial \omega}{\partial \phi}\frac{\partial \phi}{\partial k'_x} = \frac{\partial \omega}{\partial k'}\cos\phi - \frac{\partial \omega}{\partial \phi}\frac{\sin\phi}{k'} \tag{3.69}$$

$$C_{gy} = \frac{\partial \omega}{\partial k'_y} = \frac{\partial \omega}{\partial k'}\frac{\partial k'}{\partial k'_y} + \frac{\partial \omega}{\partial \phi}\frac{\partial \phi}{\partial k'_y} = \frac{\partial \omega}{\partial k'}\sin\phi + \frac{\partial \omega}{\partial \phi}\frac{\cos\phi}{k'}. \tag{3.70}$$

With the above relations the amplitude of group velocity can be expressed as:

$$c_g = \sqrt{c_{gx}^2 + c_{gy}^2} \tag{3.71}$$

Furthermore, an angle φ can be introduced ("skew angle"), as the angle between the directions of the group velocity and of the phase velocity. As mentioned before, for an isotropic material, where the direction of the group velocity coincides with the direction of the phase velocity, this angle is zero. The skew angle can be calculated as:

$$\varphi = \arctan\left(\frac{c_{gy}}{c_{gx}}\right) - \phi \tag{3.72}$$

Whereas the relation $\partial\omega/\partial k'$ can be calculated from the dispersion equations of the phase velocity, for the calculation of the term $\partial\omega/\partial\phi$ for a fixed wave number the following numerical derivation approach is proposed:

$$\frac{\partial \omega}{\partial \phi} \approx \frac{\omega(k)_{\phi+\Delta\phi} - \omega(k)_{\phi}}{\Delta\phi} \tag{3.73}$$

Assuming these relations, the group velocity dispersion can be finally obtained.

3.3.4 Shear correction factor κ

Since an accurate consideration of the contribution of transverse shear is crucial for a realistic Lamb wave solution, this section is dedicated solely to the method of shear correction factors. The correction of the transverse shear stiffness is indispensable, especially in fiber reinforced composites laminates due to the fact that, in composites, the relation between Young's modulus in fiber direction and transverse shear modulus is about one order of magnitude higher than that of homogeneous isotropic materials [Roh88]. Thus, in plate theories, the transverse shear stiffness is generally overestimated, since by the integration of the material law, the shear stiffness of a laminate is calculated by multiplying the shear modulus of each ply times the thickness. The most common method in the literature is the so-called shear correction factor method, that is to reduce the stiffness by introducing shear correction factors κ_i.

Reissner [Rei45] considered in his plate theory a shear correction factor of 5/6. Mindlin [Min51] compared natural frequencies and specified a dependency of the correction factor on the Poisson's ratio and determined a value of κ^2 of $\pi^2/12$. Uflyand [Ufl48] specified a shear correction of 2/3. Wittrick [Wit87] determined a shear correction factor of $5/(6-\nu)$ depending on the Poisson's ratio. In [Roh88] a comprehensive literature review on shear correction factors can be found. Fig. 3.5 shows shear correction factors versus Poisson's ratio as introduced by different researchers.

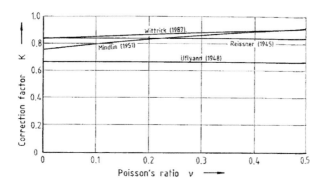

Fig. 3.5: Shear correction factors for homogeneous isotropic plates [Roh88]

Most of the approaches apply only to homogeneous, isotropic cross-sections. However, the distribution of transverse shear in a composite material with an arbitrary lay-up is not necessarily parabolic, as in an isotropic, homogeneous material. Fig. 3.6 illustrates the distribution of the transverse shear stress τ_{zx} as a consequence of transverse shear force for an isotropic plate and two composite plates with a lay-up of [90°/0°/90°] respectively [0°/90°/0°].

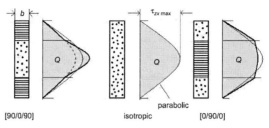

[90/0/90]	isotropic	[0/90/0]
	parabolic	

Fig. 3.6: Influence of the lay-up of a laminate on the distribution of the transverse shear stress [VDI2014]

The figure above clearly shows the influence of the lay-up and the plies properties on the distribution of the transverse shear stress, as also stated in [Whi73]. Wittrick [Wit87] even reported that it is not possible to define shear correction factors independent of the displacement mode in composites. A further sketch (Fig. 3.7) of the transverse shear distribution in a composite can be found in [Roh88].

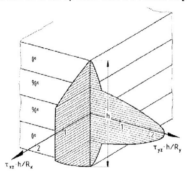

Fig. 3.7: Shear stress distribution in a composite laminate [Roh88]

As mentioned before, a common method to determine shear correction factors in composites is to match specific frequencies calculated from the plate theory to the exact 3-D solutions. However, this procedure has to be conducted for each lay-up in each direction and becomes cumbersome. Recently, Wang and Yuan [Wan07] determined shear correction factors for composite plates by matching the cut-off frequencies of A_1, S_1, SH_1, and SH_2 modes from the 3-D elasticity theory. The correction factors determined by this procedure were $\kappa_1^2 = \kappa_2^2 = \kappa_7^2 = \kappa_8^2 = \pi^2 / \left(90 - 2\sqrt{1605} \right)$, $\kappa_3^2 = \pi^2 / 12$, and $\kappa_4^2 = \kappa_5^2 = \kappa_6^2 = \pi^2 / 15$.

In this research, the factors κ_1^2 and κ_2^2 are selected according to [Sun72] as $\pi^2/12$; this was determined by matching the first flexural mode of vibration for a composite plate from the plate theory to the exact solution. κ_3^2 is chosen according to [Kan56], who determined κ_3^2 by matching the cut-off frequency of the first thickness mode to the exact 3-D solution, thus in this study the value of $\kappa_3^2 = \pi^2 / 12$ is chosen. The remaining correction factors, κ_4^2 and κ_5^2, are both chosen according to [Min59] as $\pi^2/15$.

4 Development of a software for the prediction of Lamb wave propagation in attenuative composites

4.1 Introduction

The objective of this chapter is to describe the program FIBREWAVE developed during this research. The theoretical background, on which the implementation of this software is based, was presented in chapter 3.3. The software (developed in MAT-LAB® environment) aims at predicting the dispersion and attenuation behavior of composites with arbitrary lay-ups for the Lamb wave modes A_0, SH_0, and S_0. The modes S_0 and A_0 are the most used guided wave modes in structural health monitoring applications, S_0 typically being used for crack detection and A_0 mainly for detection of delaminations. Therefore, the knowledge of the dispersion of these wave types as function of the propagation direction in composites would aid the design of effective SHM systems. Furthermore, it is important to know the attenuation of guided waves, since it determines how far the wave can propagate with a sufficient signal to noise ratio. Such attenuation models are crucial for the optimization of sensor networks in terms of location and number of sensors. Since the material damping in composites is much higher than, say, in aluminum, the significance of attenuation models is apparent.

As input data, the complex stiffness matrix C_{ij}^* of each unidirectional (UD)-ply, the lay-up and the thickness of each ply has to be given. The software defaults to pure elastic wave propagation if the user does not input any imaginary parts in the stiffness matrix. In this case, attenuation dispersion curves and attenuation values are not calculated.

The software is limited to the analysis of symmetric laminates only. This limitation is less due to the developed theoretical wave propagation model but rather due to savings in calculation time and the stability of the solution.

4.2 Program FIBREWAVE

For efficiency and stability reasons, FIBREWAVE was written using MATLAB functions (i.e. MATLAB subroutines) wherever possible. This has the advantage of clarity and structured software development. A user-friendly and self-explanatory graphical user interface (GUI) was developed. Thus, the user does not need any knowledge of MATLAB. The developed GUI is shown in Fig. 4.1.

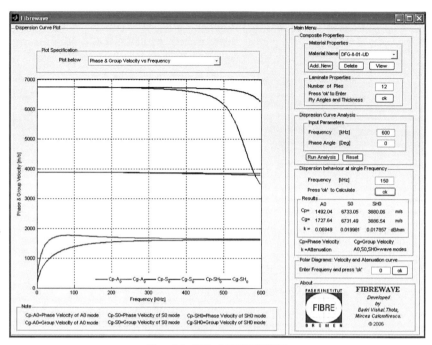

Fig. 4.1: GUI of FIBREWAVE

The input has to be done by scroll down menus. On the right hand side on the top of the GUI, the material properties have to be specified. First of all, the properties of the unidirectional plies of the analyzed laminate have to be specified. A database including different types of plies is available. Fig. 4.2 shows the input panel corresponding to the unidirectional plies used in this research (see also Tab. 5.1). However the material data base can be also extended and saved by the user.

Fig. 4.2: Input panel for mechanical UD- properties

Once the properties of the unidirectional ply in the composite laminate are given, the
next step consists in the input of the laminate lay-up. After the specification of the
number of plies, an input panel with as many cells as the specified plies appears
(Fig. 4.3). Here, the orientation and thickness of each ply has to be given.

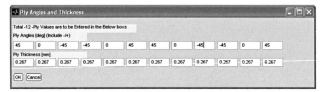

Fig. 4.3: Input panel for laminate lay-up

Finally the upper frequency limit for the output dispersion curves and the angle of
wave propagation have to be specified. The first calculation step of the software is to
obtain the properties of the laminate. As described before, this is done by the lamination theory in a subroutine called "Plies_2_laminate.m". A simplified flowchart of the
program is shown in Fig. 4.4.

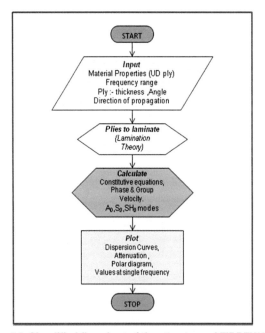

Fig. 4.4: Simplified flowchart of the structure of FIBREWAVE

4.3 Program results and discussion

The output obtained by the FIBREWAVE software is as follows:

- For a specified wave propagation angle and frequency range:
  - Dispersion curves (for both phase and group velocities) (Fig. 4.5 (a))
  - Attenuation curves (if imaginary parts of stiffness coefficients specified) (Fig. 4.5 (b))

- For a specified frequency:
  - Polar diagrams of phase and group velocities (Fig. 4.8 (top))
  - Polar diagrams of attenuation (Fig. 4.8 (bottom))

- For a specified frequency and propagation direction:
  - Numerical values of phase velocity, group velocity and attenuation (Fig. 4.9)

All results mentioned above are related to the three low-order Lamb wave modes S_0, SH_0, and A_0. Fig. 4.5 illustrates the predicted dispersion curves for an unidirectional CFRP laminate, denoted "DFG-II-01". The symmetrical S_0 Lamb mode is almost non-dispersive in the frequency range 0 – 190 kHz. The use of this wave mode for SHM applications in this frequency range may be therefore advantageous. The antisym-metric A_0 mode is highly dispersive in the low-frequency range up to a frequency of about 100 kHz and almost non-dispersive at higher frequencies.

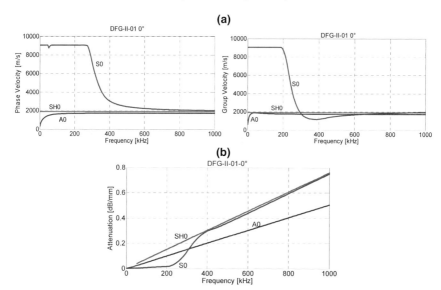

Fig. 4.5: Predicted dispersion diagrams for the laminate "DFG-II-01" in 0° direction

(a) velocity dispersion; (b) attenuation dispersion

In this research, much importance is attached to the attenuation behavior. Whereas the attenuation dispersion of A_0 and SH_0 modes increases linearly with frequency, the attenuation of S_0 shows a non-linear increase in the frequency range $260 - 360$ kHz. Large and sudden increases in attenuation can be connected to the mode shapes changing with frequency (Fig. 4.6). The rise of the S_0 attenuation in the present case can be explained as follows:

- at low frequencies (< 260 kHz), the S_0 mode is essentially compressional and possesses therefore a relatively low attenuation (s. Fig. 4.6 (a)), since the in-plane material damping of the unidirectional laminate is relatively low.
- at higher frequencies (> 260 kHz), the mode shape of S_0 gets more and more out of plane and attenuation increases drastically (s. Fig. 4.6 (a) and (b)), since the out-of-plane damping of this composite laminate increase significantly in this direction (Fig. 5.2).
- Moreover: the attenuation is related also to the phase velocity dispersion: the greater the frequential dispersion, the large the change in attenuation.

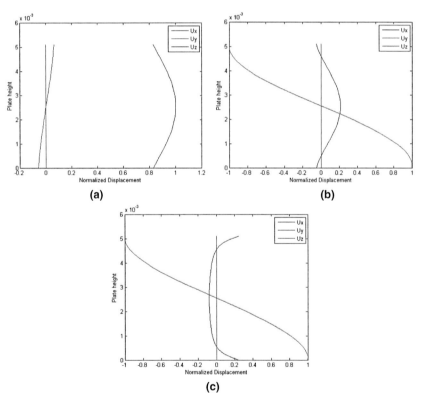

Fig. 4.6: Mode shapes of S_0 mode at different frequencies:
(a) at 260 kHz; (b) at 300 kHz; (c) at 360 kHz

In order to explain the sudden rise of the S_0 attenuation curve in Fig. 4.5b, a comparison with results provided by Srivastava and Lanza Di Scalea [Sri07] was performed (Fig. 4.7). Those attenuation diagrams were generated for many Lamb wave modes by a semi analytical finite element (SAFE) approach developed at the University of California at San Diego [Bar06].

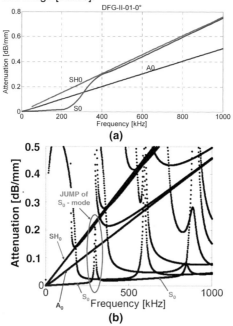

Fig. 4.7: Comparison of attenuation diagrams:

(a) provided by FIBREWAVE using HPT; (b) provided by SAFE- approach [Sri07]

The agreement between the A_0 and SH_0 mode attenuation is excellent. The attenuation of the S_0 mode agrees very well in the low frequency range (< 300 kHz) with the results obtained by the HPT. The SAFE results show a significant change/ jump in the linearity of the S_0 mode at about 300 kHz, which is in agreement with FIBRE-WAVE and HPT.

In order to obtain an understanding of the Lamb wave propagation characteristics with respect to propagation direction in an anisotropic composite laminate, polar diagrams can be output by FIBREWAVE. Fig. 4.8 shows an output of such diagrams at a fixed frequency in a typical CFRP laminate.

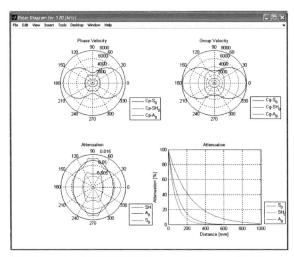

Fig. 4.8: Example of FIBREWAVE output for a specified frequency

Due to the unidirectional lay-up of the laminate analyzed in Fig. 4.8, where the anisotropy is maximal, a high dependency of wave velocity and attenuation on angle and position is observed.

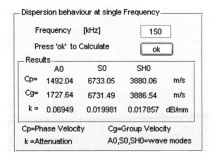

Fig. 4.9: Example of FIBREWAVE output for a specified frequency

The verification and full validation of the FIBREWAVE software is performed and reported in chapter 6.4. Additional results can be found in the appendix.

5 Non destructive material characterization

In order to verify and validate the developed mathematical model for material damping and the implemented software FIBERWAVE, six different carbon fiber reinforced plastic (CFRP) laminates were manufactured. Their viscoelastic material properties were measured non destructively by the immersion technique [Cas00]. These properties are required as input parameters in the FIBERWARE software for the validation of the predicted dispersion and attenuation behavior of the laminates.

5.1 CFRP Material

CFRP laminates were manufactured adopting the Modified Vacuum Infusion (MVI) technology at the Composite Technology Center (CTC) at Airbus Stade. This vacuum infusion process uses vacuum suction to drive resin into the laminate. The fiber material is laid dry into the mold and the vacuum is applied before resin is introduced. Once a high vacuum is achieved, resin is injected into the laminate via carefully placed tubing. CFRP laminates with different carbon fibers and lay-ups were manufactured (Tab. 5.1). The resin system for all laminates was the epoxy resin RTM6 from Hexcel Composites. The fiber volume content was $\varphi=60^{\pm4}$ % and the density of the laminate was $\rho=1.5^{\pm0.04}$ g/cm³.

Notation	Fiber type	Reinforcement type	Lay-up	Laminate thickness
DFG-II-01	HTA 6k	Unidirectional woven	$[0°]_{18}$	5.1 mm
DFG-II-02	HTA 6k	Unidirectional woven	$[0°]_{12}$	3.4 mm
DFG-II-03	HTS 12k	Non crimp fabric (Triaxial)	$[-45°/0°/45°/45°/0°/$ $-45°/-45°/0°/45°]_s$	4.7 mm
DFG-II-04	HTS 12k	Non crimp fabric (Triaxial)	$[45°/0°/-45°/-$ $45°/0°/45°]_s$	3.2 mm
DFG-II-05	HTS 12k	Non crimp fabric (Triaxial)	$[45°/0°/-45°/45°/$ $90°/45°/45°/0°/-45°]_s$	4.7 mm
DFG-II-06	HTS 12k	Non crimp fabric (Triaxial)	$[-45°/90°/45°/45°/$ $0°/-45°]_s$	3.2 mm

Tab. 5.1: CFRP laminates manufactured and tested

5.2 Immersion technique for material characterization

After considering several nondestructive material characterization techniques, it was decided to adopt the ultrasonic immersion technique developed by Hosten [Hos98] and Castaings [Cas00]. This technique is able to determine the complex 3D stiffness matrix of a composite laminate, including transversal shear and out of plane properties. The out-of-plane and transversal properties of thin-walled structures are exceedingly difficult to determine experimentally. The results of the measurements conducted in this work are reported in [Cas07].

The principle of the immersion technique consists in sending an ultrasonic plane wave through the tested sample, and capturing the whole transmitted ultrasonic field (including all echoes of longitudinal and shear waves produced inside the sample). The frequency bandwidth of the signal has to be chosen low enough such that the material is seen by the waves as homogenous and high enough to avoid lack of information about the attenuation of the waves and therefore about the imaginary parts of the complex properties. The frequency was chosen in the range 250 – 930 kHz. PZT blocks with the dimensions about 40 mm by 100 mm (Fig. 5.1) were used as transducers.

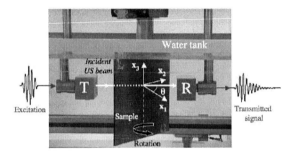

Fig. 5.1: Immersion ultrasonic rig used for measuring through-transmission ultrasonic field [Cas07]

The transmitted field is measured in a water tank for a set of angles between 0° and 50°, incremented at every 2°. These experimental data contain frequency – incident angle information, which are used to solve the inverse problem. The inverse problem consists in optimizing the complex stiffness coefficients C_{ij}^*, used as input data in a plane wave model [Hos03], so that the simulated transmitted field fits in the best way the experimental data for all angles measured in the frequency range tested.

The limitations of this technique consist in the fact that samples must have flat and parallel surfaces, such as simple plates, and must survive water immersion. Furthermore, the technique is based on the assumption that the axes of symmetry of the material are parallel (or orthogonal) to the sides of the sample. According to the arrangement of the plies, the samples are assumed to have a certain symmetry (unidirectional, hexagonal, quadratic, etc....) that will affect the number of C_{ij}^* which can be quantified. Moreover, a too large thickness of one of the plies in comparison to the other plies may also prevent good measurements because a frequency domain is chosen for which the whole sample can be considered as a homogeneous media, thus the measured C_{ij}^* coefficients are representative for the entire homogenized sample. Finally it has to be mentioned that the accuracy of the estimation of C_{ij}'' (imaginary part of C_{ij}^*) is always much worse in comparison to that of C_{ij}' (real part of C_{ij}^*); in particular the imaginary part of C_{33}^* is particularly hard to measure.

5.3 Results

From the six CFRP laminates manufactured and summarized in Tab. 5.1, only three could be characterized by the ultrasonic immersion technique. Since for the three thinner laminates denoted "*DFG-II-02*", "*DFG-II-04*", "*DFG-II-06*", the small laminate thickness prevented reliable ultrasonic measurements (essentially due to a lack of sensitivity to the material properties), these laminates could not be characterized by the immersion technique. Therefore only the remaining CFRP laminates, "*DFG-II-01*", "*DFG-II-03*", and "*DFG-II-05*", were characterized by this method.

Tab. 5.2 gives the results obtained for the unidirectional laminate "DFG-II-01". The 11-direction denotes the direction of fibers, whereas the 33-direction describes the out-of-plane direction of the laminate. As usually for this characterization technique, the relative errors for the imaginary parts of the stiffness coefficients are bigger than for the real parts. This is likely to be due to some imperfections at the interfaces between the plies, which affect more the amplitudes of the transmitted waves than their phase. And this is precisely the measurement of the amplitudes, which are determinant for the accuracy of the imaginary parts.

[GPa]	C_{11}	C_{22}	C_{33}	C_{12}	C_{13}	C_{23}	C_{44}	C_{55}	C_{66}
C'_{ij}	125 ± 8	13.9 ± 0.4	14.5 ± 0.4	6.3 ± 0.6	5.4 ± 0.4	7.1 ± 0.2	3.7 ± 0.3	5.4 ± 0.4	5.4 ± 0.6
C''_{ij}	3 ± 1	0.6 ± 0.2	0.6 ± 0.2	0.9 ± 0.6	0.4 ± 0.2	0.23 ± 0.15	0.12 ± 0.06	0.3 ± 0.15	0.5 ± 0.25

Tab. 5.2: Results of C^*_{ij} for the CFRP laminate "DFG-II-01" with $[0°]_{18}$

The measured values (Tab. 5.2) indicate that the plane 2-3 is isotropic ($C^*_{22} \approx C^*_{33}$ and $C^*_{44} \approx (C^*_{33} - C^*_{23})/2$), as expected for an unidirectional laminate. The high values of C^*_{33} indicate high stiffness in the fiber direction. The imaginary part of C^*_{11} is particularly hard to measure with the immersion technique; the first estimation during the measurements was $C''_{11} = 10^{\pm 5} \, GPa$. However, compared to measurements reported in the literature, e.g. in [Han92], this value was unusually high, in particular when one considers that the attenuation in fiber direction should be smaller than in any other direction. Thus, different additional measurements and optimization approaches were adopted allowing C''_{11} to be adjusted to $C''_{11} = 3^{\pm 1} \, GPa$. As damping factors, the ratio [Han92] between the imaginary part and the real part of the stiffness moduli ($d_{ij} = E''_{ij}/E'_{ij}$, resp. $d_{ij} = G''_{ij}/G'_{ij}$) were calculated. The damping factors d_{11}, d_{22}, d_{33} denote the extensional damping and the coefficients d_{23}, d_{13}, d_{12} the shear damping (Fig. 5.2). However, these values have to be handled with care, if one keeps in mind the error interval of the determined imaginary parts of C^*_{ij} .

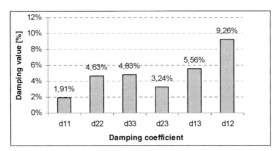

Fig. 5.2: Comparison of damping factors for the unidirectional laminate "*DFG-II-01*"

As shown in Fig. 5.2, the lowest material damping factor is in the direction of the fibers, since the matrix, with its excellent damping properties, plays a minor role in this direction. Considering just the matrix alone, its shear damping properties are much higher than the material damping properties for extensional loading. Considering the interaction between matrix and fibers, the high damping factor d_{12} (the highest damping factor for the laminate "DFG-II-01") could be explained. Finally, it can be stated that the measured values seem to be consistent. (To the knowledge of the author of this research there is no other nondestructive characterization method available that enables the measurement of the complex 3D stiffness matrix for composites).

The properties of the two other CFRP laminates characterized by the immersion technique are given in Tab. 5.3 and Tab. 5.4.

[GPa]	C_{11}	C_{22}	C_{33}	C_{12}	C_{13}	C_{23}	C_{44}	C_{55}	C_{66}
C'_{ij}	70 ± 0.4	33 ± 0.5	14.7 ± 0.1	23.9 ± 1	6.2 ± 0.2	6.8 ± 0.3	4.2 ± 0.1	4.7 ± 0.1	21.9 ± 1
C''_{ij}	1.8 ± 0.5	1.4 ± 0.5	0.5 ± 0.1	0.9 ± 0.5	0.3 ± 0.2	0.2 ± 0.1	0.17 ± 0.05	0.2 ± 0.03	0.5 ± 0.3

Tab. 5.3: Results of C^*_{ij} for the CFRP laminate "DFG-II-03"

with [-45°/0°/45°/45°/0°/-45°/-45°/0°/45°]$_s$

[GPa]	C_{11}	C_{22}	C_{33}	C_{12}	C_{13}	C_{23}	C_{44}	C_{55}	C_{66}
C'_{ij}	56.5 ± 1.5	44.5 ± 1.5	14.7 ± 0.3	25.5 ± 3	6.4 ± 0.4	6.7 ± 0.5	4.4 ± 0.1	4.5 ± 0.1	22.1 ± 3
C''_{ij}	2.3 ± 1	3 ± 1	0.5 ± 0.2	1.8 ± 1	0.1 ± 0.08	0.6 ± 0.4	0.2 ± 0.1	0.2 ± 0.07	1.5 ± 1

Tab. 5.4: Results of C^*_{ij} for the CFRP laminate "DFG-II-05"

with [45°/0°/-45°/45°/90°/45°/45°/0°/-45°]$_s$

As already mentioned, the nondestructive material characterization by the immersion technique could not be achieved for the other three remaining CFRP laminates. Therefore, we decided to analyze if the laminate properties can be estimated from the unidirectional ply properties. In order to evaluate the errors of this approach, first the C^*_{ij} properties of laminate "DFG-II-03" are calculated by the assumption that all unidirectional plies of this laminate have the properties of the unidirectional laminate "DFG-II-01". These properties were taken and transformed according to the lay-up of laminate "DFG-II-03". In this way an estimation of the properties of any laminate containing the same UD-plies can be achieved. The results for laminate "DFG-II-03" are shown in Tab. 5.5 in comparison to the values obtained by the immersion technique.

Experimentally obtained by the immersion technique								
[GPa]	C_{11}	C_{22}	C_{33}	C_{12}	C_{13}	C_{23}	C_{44}	C_{55}	C_{66}
C'$_{ij}$	70	33	14.7	23.9	6.2	6.8	4.2	4.7	21.9
C"$_{ij}$	1.8	1.4	0.5	0.9	0.3	0.2	0.17	0.2	0.5
Theoretically obtained by lamination theory using UD ply properties from "DFG-II-01"								
[GPa]	C_{11}	C_{22}	C_{33}	C_{12}	C_{13}	C_{23}	C_{44}	C_{55}	C_{66}
C'$_{ij}$	70	33	14.5	23.75	5.97	6.53	4.27	4.83	22.85
C"$_{ij}$	2.23	1.43	0.6	0.87	0.34	0.29	0.18	0.24	0.47

Tab. 5.5: Comparison of C$^{*}_{ij}$ for the CFRP laminate "DFG-II-03"

with [-45°/0°/45°/45°/0°/-45°/-45°/0°/45°]$_s$ obtained in different ways

As Tab. 5.5 shows, the C$^{*}_{ij}$ coefficients for a laminate with a arbitrary lay-up can be estimated using the C$^{*}_{ij}$ coefficients of the UD- plies within the same range of error as the immersion technique is able to detect them.

5.4 Lamb Wave attenuation factors

In this section, the computation of Lamb wave attenuation factors was performed and summarized for all the CFRP laminates tested in this study. For those laminates, for which the "immersion technique" could not provide reliable results, the properties of the laminates were approximated by the lamination theory. Based on the laminate properties, the attenuation factors were computed by the developed FIBREWAVE software for all six laminates studied in this research. In Tab. 5.6, the results for Lamb waves at a single center frequency of 300 kHz are summarized.

	Attenuation factor k" [dB/mm] @ 300 kHz								
	S_0 mode			SH_0 mode			A_0 mode		
Laminate	0°	90°	45°	0°	90°	45°	0°	90°	45°
DFG-II-01	0.0249	0.0749	0.0376	0.2298	0.2313	0.0876	0.1518	0.1075	0.1468
DFG-II-02	0.0249	0.0753	0.0373	0.2287	0.2301	0.0877	0.1503	0.1080	0.1545
DFG-II-03	0.0399	0.0716	0.0393	0.0358	0.0353	0.0688	0.1430	0.1316	0.1373
DFG-II-04	0.0403	0.0724	0.0692	0.0359	0.0361	0.0398	0.1420	0.1331	0.1365
DFG-II-05	0.0464	0.0529	0.0363	0.0365	0.0412	0.0644	0.1389	0.1365	0.1412
DFG-II-06	0.0524	0.0524	0.0408	0.0362	0.0355	0.0665	0.1386	0.1369	0.1359

Tab. 5.6: Summary of theoretical attenuation factors at 300 kHz

Fig. 5.3 shows the theoretical attenuation factors for the unidirectional CFRP laminate DFG-II-02. As expected, the attenuation of the S_0 mode in 0° (fiber direction) is much smaller than the attenuation of the same mode in 90°. This behavior can be explained by the fact that the matrix properties, which are much more attenuative, predominate in 90° direction. However, comparing the attenuation of the A_0 mode at 0° and 90°, the behavior is not analog, since the transversal shear properties are crucial for this Lamb wave mode at this frequency, and they did not differ significantly as function of the propagation direction.

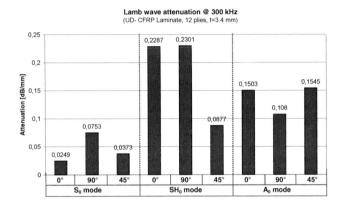

Fig. 5.3: Lamb wave attenuation at 300 kHz in the laminate "DFG-II-02"

48

6   Experimental investigations of wave propagation (wave excitation by PWAS)

6.1   Experimental set-up

The experimental set-up for Lamb-wave propagation consists of the four-channel digital oscilloscope Tektronix TDS5034B, the arbitrary signal generator HP33120A and the wideband amplifier Krohn-Hite 7602. The set-up is illustrated in Fig. 6.1. For actuating and sensing purposes, round piezoelectric wafer active sensors (PWAS), with a thickness of 0.2 mm and a diameter of 7 mm, were used. The PWAS were bonded on the laminate surface with M-Bond 200 adhesive. The bonding procedure followed Vishay Micro-Measurements Instruction Bulletin B-127-14 [Vis05] proposed initially for strain gauge installation. The PWAS were connected with thin insulated wires AWG 38 with a diameter of 0.1 mm. The piezoelectric material of the PWAS was the ceramic Lead Zirconate Titanate (PZT) from American Piezo Ceramics Inc. with the notation APC-850. The mechanical and piezoelectric properties of the APC-850 material were given earlier in Tab. 2.1. The signal generator was used to apply an excitation to PWAS#1 (Fig. 6.1), whereas the remaining PWAS were used as sensors, to collect the excited wave packets. Since the HP33120A has not enough output power, the Krohn-Hite 7602 amplifier was used to increase the amplitude of the excitation voltage.

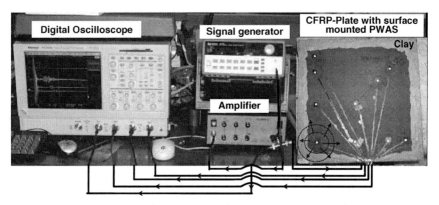

Fig. 6.1: Experimental set-up for Lamb wave propagation

Fig. 6.2 shows a sketch of the experimental setup with dimensions and locations of all ten applied PWAS. Since a four channel oscilloscope was used, only the signals of the PWAS actuator and three PWAS receivers could be collected at the same time. The CFRP laminates used were flat panels with the overall dimensions of 500 x 500 mm of lay-up and properties corresponding to Tab. 5.1 to Tab. 5.5.

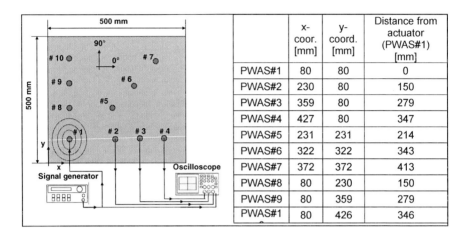

	x-coor. [mm]	y-coord. [mm]	Distance from actuator (PWAS#1) [mm]
PWAS#1	80	80	0
PWAS#2	230	80	150
PWAS#3	359	80	279
PWAS#4	427	80	347
PWAS#5	231	231	214
PWAS#6	322	322	343
PWAS#7	372	372	413
PWAS#8	80	230	150
PWAS#9	80	359	279
PWAS#1	80	426	346

Fig. 6.2: Experimental set-up for Lamb wave propagation

6.2   Excitation of Lamb waves

As already mentioned in chapter 2.3, PWAS are strain coupled (through the adhesive shear-layer) transducers. The used circular-shaped PWAS excite omni-directional Lamb waves that propagate in circular wave fronts.

It is not difficult to construct a continuous single-frequency excitation signal. However if a continuous excitation is applied, it is not possible to assign the collected signals to the excited ones as needed here. The time of flight of a signal has to be deter-mined in order to calculate its propagation velocity and the ratio of the amplitudes is required in order to determine the attenuation. Therefore, an excitation signal of small and limited duration, a so called "tone burst" is needed. However, it is difficult to ex-cite single-frequency signals with tone burst signals, since the sharp transition at the start and end of such a signal excites additional frequencies, which is undesirable. In particular, for the excitation of dispersive Lamb waves, the excitation of coherent sin-gle-frequency waves is very important. A tone burst signal filtered through a Hanning window was used as excitation signal, as proposed by Giurgiutiu [Giu04].

The excitation function can be thus expressed as

$$u(t) = u_0 \sin(2\pi f t)\frac{1}{2}\left[1 - \cos\left(\frac{2\pi f t}{n}\right)\right], \quad t \in \left[0, \frac{n}{f}\right] \tag{6.1}$$

where n is the number of counts and f is the central frequency. In this research three count signals were used with a voltage of 30 V peak to peak. For a center frequency of $f = 100$ kHz the excitation signal is illustrated in Fig. 6.3. The signal in the time do-main is shown in Fig. 6.3a. The time-frequency domain signal, obtained by the con-tinuous wavelet transform (CWT) with the Gabor function as mother wavelet, is shown in Fig. 6.3b.

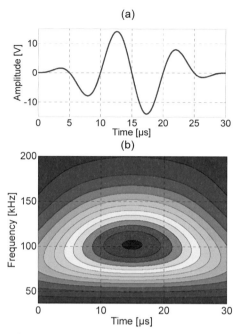

Fig. 6.3: Burst signal used for the excitation of Lamb waves (here f= 100 kHz):

(a) in time domain; (b) in time-frequency domain (CWT)

The center frequency of 100 kHz can be clearly seen in the CWT-spectrogram, however it is evident that many other frequencies (30 - 200 kHz) around the center frequency are excited, but of course of lesser amplitude.

The excitation signal explained above is numerically digitalized and sent from the PC through an USB - GPIB interface to the signal generator where is stored permanently.

6.3   Excitability

According to Giurgiutiu [Giu04] the excitation of Lamb waves by surface mounted PWAS transducers has a strong amplitude frequency dependency. This excitability was investigated during the first part of the experiments. Giurgiutiu and his co-workers reported in [Giu04] and [Bot05] that, for certain excitation frequencies the excitation of A_0 modes is much stronger, whereas for other excitation frequencies the excitation of A_0 is completely rejected, whereas the S_0 excitation is very strong. In [Giu04], it was analytically shown that maximum Lamb wave excitation occurs when the effective PWAS length is an odd-integer multiple of the half wavelength of the excited Lamb wave mode.

Thus, the relation for maximum excitation can be expressed as:

$$\frac{\lambda}{2} = \frac{l_{eff}}{(2m+1)} \qquad m = 0, 1, 2, 3..... \qquad (6.2)$$

where λ is the wavelength of the excited Lamb wave mode and l_{eff} denotes the effective length of the PWAS transducer. Of course this principle applies also for a PWAS receiver used for sensing applications, which means that such a receiver is able to selectively detect Lamb wave modes. Raghavan and Cesnik [Rag04] extended the approach of Giurgiutiu [Giu04] from plane waves to circular crested Lamb waves.

During the experimental tests, it was observed that multiple wave reflections from the plate boundaries occur, which make the analysis difficult and in many cases even impossible. The reflections and the initial waves superposed, thus it was not possible to clearly distinguish them. To stop reflections, a layer of modeling clay with a width of about 30 mm was applied on all laminates, as illustrated in Fig. 6.1, right. In Fig. 6.4, the waves collected by PWAS#2 (
Fig. 6.2) on the unidirectional CFRP laminate "DFG-II-01" with and without clay are illustrated. The envelope was obtained by the Hilbert transform. Since the effect of clay of reducing reflections from panel edges is obvious, a layer of modeling clay was applied to the laminates, for all experimental tests conducted in this research.

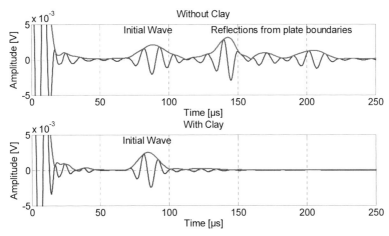

Fig. 6.4: Collected Lamb waves at PWAS#2 on the laminate "DFG-II-01" with and without clay at plate edges (here shown f_c=100 kHz)

Experimental pitch-catch results of excitability and detection of the Lamb wave modes A_0, SH_0 and S_0 are presented next. Lamb waves were excited from 10 kHz to 400 kHz in steps of 10 kHz. As already explained, in this frequency range no other guided wave modes than the antisymmetrical A_0 wave, the shear horizontal SH_0 and the symmetrical S_0 exist. However, not all three modes can be excited and detected over the entire frequency range from 10 kHz to 400 kHz by the PWAS.

In Fig. 6.5, the excitability results corresponding to the unidirectional laminate "*DFG-II-02*" are presented.

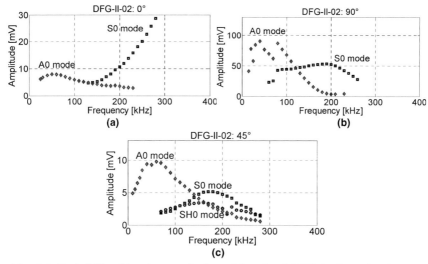

Fig. 6.5: Excitability of Lamb waves in the unidirectional CFRP laminate "*DFG-II-02*":

(a) in 0° direction; (b) in 90° direction; (c) in 45° direction

The amplitudes illustrated in Fig. 6.5 are obtained by PWAS#2 for the 0° direction, PWAS#5 for the 45° wave propagation direction, and by PWAS#8 for the 90° direction. As mentioned before, PWAS#1 was used as actuator in all experiments. The excitability on all three propagation angle studied showed the same tendency: at lower frequencies (< ~130 kHz) the A_0 mode is much stronger excited. The maximum excitation of the A_0 mode appears at a frequency of 55-60 kHz and is almost independent of the propagation direction. This is due to the fact, that the A_0 wavelength is very similar in all directions. The SH_0 mode cannot be excited (and sensed) in 0° and 90° directions, but was very well excited in 45° direction. However, for strongly anisotropic laminate (unidirectional laminate) it was clearly observed that the maximum of the S_0 mode excitation does not occur at the same frequency for all directions. This can be traced back to the fact that the wavelength of the S_0 mode at a given frequency is very depended on direction due to the much higher phase velocity on fiber direction and much lower velocity orthogonal to the fibers. This difference is not as significant for the A_0 mode. In conclusion, it can be summarized that at lower frequencies below about 130 kHz the excitation of the A_0 mode predominates, whereas at higher frequencies (130 – 300 kHz) the S_0 is the predominant mode excited. For SHM application, these frequencies, at which one mode is excited very strongly, are important, since all the other modes are rejected. This makes the interpretation and the analysis of the wave patterns much easier and the post-processing more reliable.

Further excitability studies performed on multiaxial CFRP laminates with different layups are presented next. Fig. 6.6 shows the excitability plots for the multiaxial laminate with the notation "*DFG-II-05*", which is a laminate with a lay-up of [45°/0°/-45°/45°/90°/45°/45°/0°/-45°]$_s$.

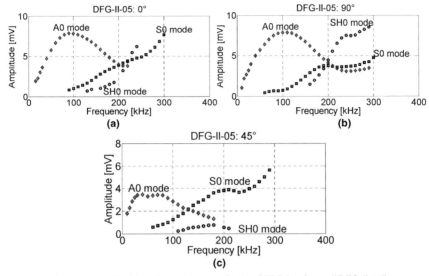

Fig. 6.6: Excitability of Lamb waves in the CFRP laminate "*DFG-II-05*":

(a) in 0° direction @ PWAS#2; (b) in 90° direction @ PWAS#8; (c) in 45° direction @ PWAS#5

The excitability diagrams of this multidirectional CFRP laminate show a behavior similar to that already observed, for the unidirectional laminate: at lower frequencies the A_0 mode is predominant, at higher frequencies S_0 and SH_0 can be excited much stronger.

Excitability plots of all CFRP laminates studied in this thesis are presented in the Appendix.

6.4   Verification of the developed theory and software

The main objective of the wave propagation experiments conducted and presented in this chapter was to verify and validate the developed complex higher order plate theory for Lamb wave propagation and to prove its implementation into the developed software FIBREWAVE presented in chapter 4. Predicted and experimental dispersion curves and attenuation dispersion curves for the six different CFRP laminates in three propagation directions (0°, 45° and 90°) are compared next.

Dispersion curves are diagrams showing the variation of wave propagation velocity (phase or group velocity) with frequency. In order to determine experimentally the dispersion curves, the arrival time of each wave mode has to be measured. Once the arrival time is known, the time of flight t (TOF) can be calculated from the difference of the arrival time and the time at which the signal started. Once the distance d is known, the wave velocity c for each wave mode can be simply obtained by the relation $c = d/t$. As illustrated in Fig. 6.7, the maximum of the envelope obtained by the Hilbert transform was used to define the arrival time. To obtain the TOF at each PWAS and for each guided wave mode, the time, which denotes the maximum of the envelope of the excitation signal was subtracted from the arrival time.

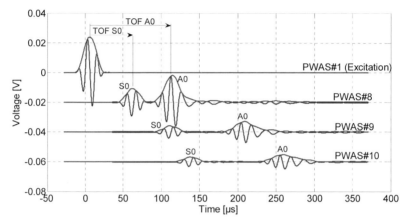

Fig. 6.7: Wave propagation in laminate "DFG-II-02" and definition of TOF

(in 90° direction at excitation frequenc y 80 kHz)

Fig. 6.7 shows the excitability of Lamb waves: at 80 kHz excitation center frequency, both A_0 and S_0 Lamb modes are excitable. However the excited amplitude of A_0 is higher than that of S_0 (compared also to Fig. 6.5 (b)). The group velocities obtained from Fig. 6.7 are very consistent as shown in Tab. 6.1.

	Group velocity [m/s]	
	S_0 mode	A_0 mode
@ PWAS#8	2623	1388
@ PWAS#9	2614	1382
@ PWAS#10	2614	1379

Tab. 6.1: Experimentally obtained wave velocities in laminate "DFG-II-02" in

in 90° direction at 80 kHz

However the time domain representations are not able to give any information on the frequency content of the signals. To achieve this, the continuous wavelet transform (CWT) was applied to the signals from Fig. 6.7 using the Gabor function as mother wavelet. The contour plots are shown in Fig. 6.8.

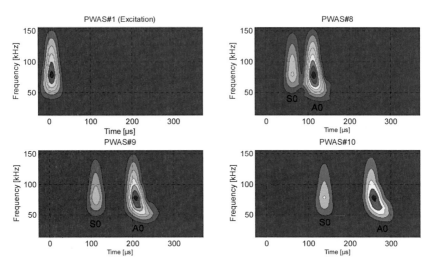

Fig. 6.8: Contour plot of CWT, laminate "DFG-II-02", frequency 80 kHz, 90°- direction

As expected, the highest magnitudes are in the frequency range around 80 kHz, however although with much less amplitudes, it can also be observed that frequencies between about 40 kHz to 150 kHz are present in the signal. This is predominantly due to the fact (explained in chapter 6.2) that it is not possible to excite a perfect single frequency waves with any transient tone-burst signal. The arrival of the wave modes S_0 and A_0 and the fact, that the amplitude of the A_0 mode is much higher than that of S_0 can be seen clearly in Fig. 6.8. It can be also clearly seen, that the CWT contour plot corresponding to the A_0 mode is not as "straight" as for the S_0 mode, but it is, slightly "curved" in the lower frequency range below ca. 80 kHz. This behavior is due to the strong dispersion behavior of the A_0 mode in the lower frequency range and means that the lower frequency components arrived a little later than the higher frequency components, which arrived almost at the same time. Thus, a CWT contour plot can be used to detect the type of Lamb wave mode.

The diagrams illustrated in Fig. 6.9 and Fig. 6.10 show the dispersion behavior of the unidirectional laminate "DFG-II-0

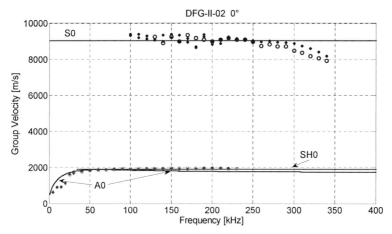

Fig. 6.9: Dispersion diagram for the laminate "DFG-II-02" in 0° direction

In the frequency range 0 – 400 kHz, the S_0 mode is non dispersive (Fig. 6.9), with a theoretical group velocity of 9034 m/s in the 0° direction, which denotes the direction of fibers in the unidirectional laminate "DFG-II-02". Fig. 6.10 shows the velocity dispersion diagram of the same unidirectional laminate (DFG-II-02), but for wave propagation in 90° direction. The most significant change is in the propagation velocity of the S_0 mode, which drops from about 9000 m/s in the low frequency range to about 2600 m/s. In both diagrams the SH0 mode could not be excited or detected experimentally, thus it could not be compared with theoretical results.

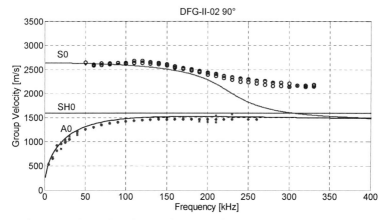

Fig. 6.10: Dispersion diagram for the laminate "DFG-II-02" in 90° direction

A further comparison of theoretical and experimental dispersion data are shown in Fig. 6.11 for the case of the CFRP laminate "DFG-II-04" in 0° laminate direction. In this case, all three guided wave modes (A_0, SH_0, S_0) could be excited. As in most of the studies performed in this dissertation, the agreement between theoretical and experimental results is best for the A_0 mode.

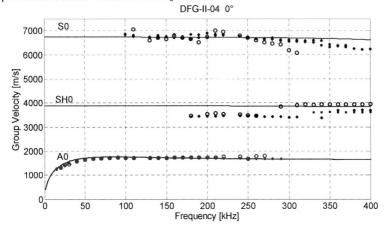

Fig. 6.11: Dispersion diagram for the laminate "DFG-II-04" in 0° direction

7 Experimental investigation of wave propagation under impact hammer excitation

In this chapter, experimental investigations for a passive impact sensing system are presented, which requires only sensors to "listen" to the structure, but no actuators. Although in this work for the passive sensing piezoelectric sensors are used, it is theoretically possible to use any other sensors, e.g. strain gages or optical fiber sensors, to collect strain-history signals from the structure provided as long as the sensors possess the required sensitivity and signal to noise ratio.

7.1 Experimental set-up

A hand-held impulse force hammer model 086C03 from PCB Piezotronics was used to impact the structure. A piezoelectric force transducer was installed in the tip of the hammer, to record the contact force history. The measured output voltage can be simply converted into force by the relation:

$$Force[N] = \frac{Voltage[V]}{0.00231[V/N]} \tag{7.1}$$

The same PWAS as already applied for the active wave excitation in chapter 6 were used. The only difference in comparison to the experimental set-up for the PWAS excitation in chapter 6 is that a different digital oscilloscope, the 4-channel Tektronix TDS3014B, and an impact hammer, instead of the active Lamb wave excitation by one of the surface bonded PWAS were used. Fig. 7.1 shows the experimental set-up for impact investigations.

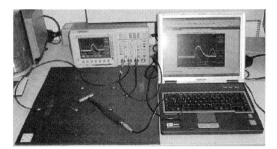

Fig. 7.1: Experimental Set-up for wave propagation due to impact

Since the same CFRP samples as in chapter 6 were used, the arrangement of the PWAS is exactly the same.

7.2 Measurements and Signal Processing

Fig. 7.2 illustrates an analyzed impact case, in which a low-energy impact (applied by the hand-held hammer) hits the CFRP laminate at the marked position. Additionally to the impact force history, the strain-history signals at PWAS#8, PWAS#9 and PWAS#10 were recorded. The measured voltage generated by the PWAS was converted into the sum of the in-plane strains by the relations presented in chapter 2.3. The scope was to record the propagation of wave excited by the impact in only one laminate direction, i.e., in the propagation direction perpendicular to the fibers of the unidirectional laminate DFG-II-01. Fig. 7.3 shows the recorded signals.

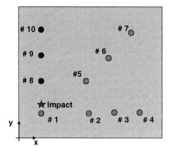

	x-coor. [mm]	y-coord. [mm]	Distance from impact location [mm]
Impact	80	180	0
PWAS	80	230	50
PWAS	80	359	179
PWAS	80	426	246

Fig. 7.2: Impact force history and measured strain-histories at different PWAS

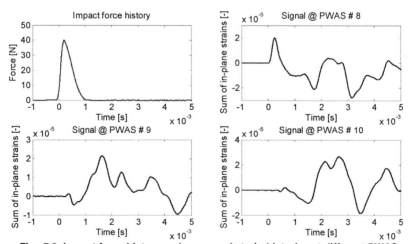

Fig. 7.3: Impact force history and measured strain-histories at different PWAS

As shown in Fig. 7.3, the contact time of the impact is in the range of 1 ms and the maximum force peak amounts to about 40 N. For such low-energy impacts no damages were expected in the tested CFRP- laminates. Since the waves recorded by the PWAS were excited (in contrast to the in-plane PWAS excitation) by an out of plane, transversal impact, it is supposed that A_0 Lamb mode will predominate. However, as mentioned before, A_0 Lamb modes are highly dispersive in the low frequency range, which might complicate the time of flight detection for a broadband excitation. In order to examine the impact force history and the recorded wave signals in the time-frequency domain, contour plots of the four signals using the CWT with the Gabor function as mother wavelet were performed (Fig. 7.4). As shown in Fig. 7.4, the frequency content in the signals is below 3 kHz, whereas the strongest signals are even below 1 kHz.

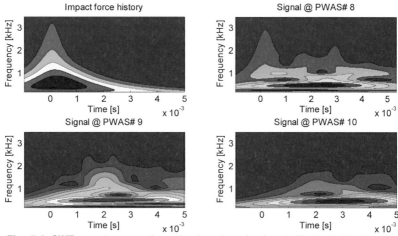

Fig. 7.4: CWT- spectrograms from the time-domain signals illustrated in Fig. 7.3

The aim of this impact experiments is to provide a basis for the development of a method for impact detection (localization of impact position and determination of impact force histories). For the localization of the impact arrival time, information at each PWAS is necessary. As shown in Fig. 7.3, the detection of arrival times is a challenging task. It is not possible to ensure that at all PWAS the arrival time detected belongs to the same signal frequency. Since the waves excited by impacts are highly dispersive, it can be expected that a big error will creep in. In our work, the magnitude of the wavelet (WT) coefficients is investigated. Fig. 7.5 shows the magnitude of WT coefficients vs. time for a scale of 667, (which corresponds to a frequency of 1.5 kHz for the sampling frequency of 1 MHz, used here). The reason for choosing such a low frequency is that the highest amplitude in the signals excited by the impact hammer occurs at this frequency, thus the signal is strong and can be clearly distinguished from noise. However, the disadvantage of choosing this frequency is that the flexural waves are highly dispersive in this frequency range.

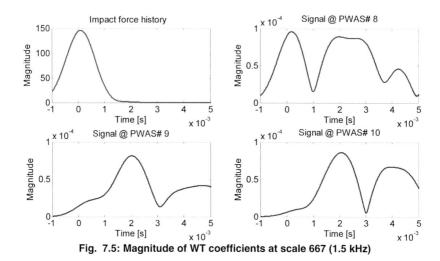

Fig. 7.5: Magnitude of WT coefficients at scale 667 (1.5 kHz)

8   Numerical studies with explicit finite element analysis

In order to obtain a better understanding and to visualize the propagation of waves in anisotropic composites, finite element analyses were also performed. For the numerical simulation of the wave propagation, an explicit finite element code is used, since short-time, transient wave cannot be computed by implicit finite element methods. The main solution in explicit codes is based on explicit time integration to study nonlinear dynamic problems. In such codes, typically the central difference method is used for the numerical time integration. The integration method is stable only when the calculation time-step is smaller than the critical time-step, which is dependent on the highest frequency in the system. The relation can be then expressed as:

$$\Delta t \leq \Delta t_{crit} = \frac{2}{\omega_{max}}$$ (8.1)

The commercial software LS-DYNA was used for the finite element analysis. LS-DYNA is a general purpose finite element code for analyzing the dynamic response of structures [Hal98]. Fig. 8.1 shows a typical computation cycle of LS-DYNA, which is however very similar to other explicit finite element codes.

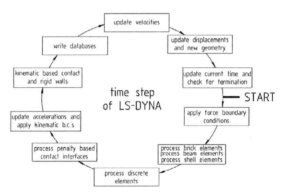

Fig. 8.1: Computation cycle of explicit FEA [Rei04]

The version of LS-DYNA available for this research was not able to take into account anisotropic material damping. Since, as found in this dissertation, the influence of guided wave attenuation is not negligible particularly in the far field and for waves at higher frequencies (30 -500 kHz), it was decided in this chapter to conduct investigation on flexural waves after an impact in the near field only, where the anisotropic material damping plays a minor role. Experimentally, it was found that a low velocity impact event does not excite flexural waves with significant components of frequency above 30 kHz.

8.1   Finite element model

In this subsection, the finite element modeling of an impact event on a CFRP laminate is described. The dimensions of the FEA model correspond to the geometrical properties adopted during the experimental tests. As mentioned before, a 500 x 500 CFRP laminate plate was investigated. The finite element discretization was performed with shell elements. In order to avoid Hourglass oscillations, fully integrated 4-node shell elements (ELFORM 16) were used. (Hourglass modes are non-physical zero energy deformation modes, which often appear in explicit finite element codes.)

In order to consider fiber reinforced laminates, the material model MAT#54 (*MAT_ENHANCED_COMPOSITE_DAMAGE) was adopted. This material model uses laminated shell theory based on Mindlin's theory [Min51] and includes transverse shear deformations. It was decided to use this material model, since it has very interesting degradation algorithms for composite materials. In this material model, the material properties of each unidirectional ply has to be entered and the lay-up of the laminate. The LS-DYNA material input card corresponding to the CFRP laminates tested in this research is given in Fig. 8.2.

```
*MAT_ENHANCED_COMPOSITE_DAMAGE
$^MAT0001
$       MID         RO         EA         EB         EC       PRBA       PRCA       PRCB
       10.00000148      121.7       10.3       10.3      0.029      0.018       0.50
$       GAB        GBC        GCA         KF       AOPT
        5.4        3.7        5.4        0.0        2.0
$     BLANK      BLANK      BLANK         A1         A2         A3     MANGLE
          0          0          0        1.0        0.0        0.0        0.0
$        V1         V2         V3         D1         D2         D3     DFAILM     DFAILS
        0.0        0.0        0.0        0.0        1.0        0.0        0.0        0.0
$     TFAIL       ALPH       SOFT       FBRT      YCFAC     DFAILT     DFAILC        EFS
        0.0        0.0        0.0        0.0        0.0        0.0        0.0        0.0
$        XC         XT         YC         YT         SC       CRIT       BETA
        0.90        1.5      0.200      0.080      0.048       54.0        0.0
```

Fig. 8.2: Input deck for FEA of an unidirectional ply of the CFRP laminate "DFG-II-01"

The total time of calculation was set to 3 ms. The analyzed 500 x 500 mm CFRP plate was discretized using 10000 shell elements, with the dimensions 5 x 5 mm. It is important to obtain at least 15-20 nodes per smallest wavelength. The impact load was applied with respect to time and equally distributed on the corresponding nodes in the region of impact. In order to allow the comparison and validation with experimental results, the impact force histories obtained from experimental measurements were used.

8.2   Results

Different impact configurations at several positions and on several CFRP laminates were analyzed by FEA and compared to the experimental tests. In this subsection the impact illustrated in Fig. 8.3 is discussed.

	x-coor. [mm]	y-coord. [mm]	Distance from impact location [mm]
Impact	180	80	0
PWAS#2	230	80	50
PWAS#3	359	80	79
PWAS#4	427	80	247

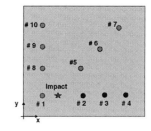

Fig. 8.3: Impact on the CFRP laminate "DFG-II-01"

As impact loading, in the FEA the original impact force-time history measured experimentally was used (Fig. 8.4 (left)). The nine nodes on which the force-time history was applied as transient nodal forces are shown on the right hand side of Fig. 8.4.

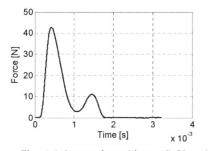

Fig. 8.4: Impact force history (left) and FE-nodes on which it was applied (right)

The computational time for the transient FE- analysis with a calculation time step of 0.49 µs and a total calculated time of 3 ms is about 15 minutes on a two CPU, personal computer.

For the (qualitative) evaluation and visualization of the wave propagation during and after the impact loading, the strains in x- direction at the upper surface of the laminate are plotted in Fig. 8.5 at different times. The impact position as well as the locations of the PWAS are indicated in the first plot at t= 0.3 ms. Moreover, the plots show that first reflections at the boundaries of the laminated plate occurred already at t= 0.4 ms. At this time, the impact loading is not yet finished (cf. Fig. 8.4 (left)).

The out of plane displacements of the impact loaded CFRP plate at different time steps were analyzed in Fig. 8.6. The figures clearly show the strong anisotropic material behavior of the analyzed CFRP laminate and the higher propagation velocity in the direction of the fibers (x-direction).

However, neither Fig. 8.5 nor Fig. 8.6 can give any information about the wave modes excited by the impact. As mentioned in one of the previous chapters in this dissertation, it can be assumed that such a transversal impact excites predominantly the A_0 mode (respectively flexural waves in this lower frequency range). Information on the wave mode is given in particular by the strain distribution in the cross section of the laminate. The wave mode can be extracted from the shape of the strain distribution, since the mode shapes of the different wave types are known. Fig. 8.7 shows the x- strain at three positions across the laminate thickness (at upper surface, mid surface and lower surface) for a shell element situated closely the PWAS # 3. The x-strain at the upper surface is almost identical in terms of magnitude, but with a reversed sign. The mid surface strain is almost zero, which indicates that there is a flexural wave excited predominately by the impact event. The FEA results also showed that surface bonded PWAS do not influence the wave propagation due to an impact event. This is to be expected, since the waves excited by an impact have relatively high wavelengths, which do not interact with small imperfections.

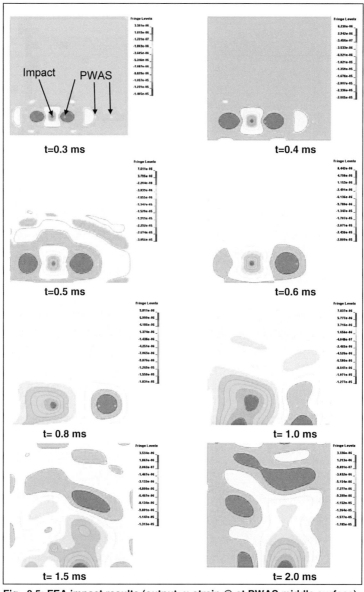

Fig. 8.5: FEA impact results (output: x-strain @ at PWAS middle surface)

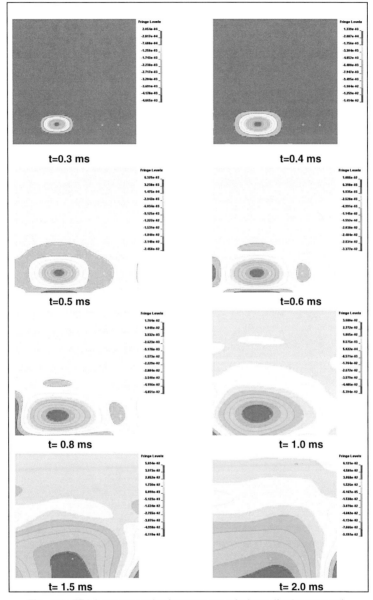

Fig. 8.6: FEA impact results (output: out-of-plane displacements)

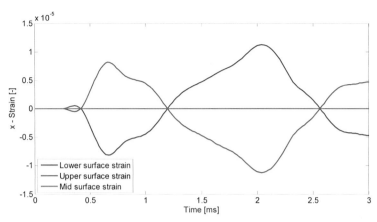

Fig. 8.7: Strain distribution across the thickness of the laminate near PWAS# 3 (FEA results)

8.3   Experimental vs. numerical results

The goal of this section is to check the agreement between the explicit FEM analysis and the experimental wave propagation results and quantify the accuracy. Since the voltage output of the PWAS utilized is proportional to the sum of in-plane strains, these strains were computed numerically and then compared with PWAS data collected at three positions on the CFRP plate. Fig. 8.8 illustrates this comparison, attesting an excellent agreement at the early stage of the wave propagation i.e., for times < 0.5 ms. The experimental and FEA curves are almost congruent in this time range. The agreement is still satisfactory for the time range between 0.5 and 1.3 ms both qualitatively and quantitatively. However, for the late stage of the wave propagation i.e., for time > 1.5 ms, there is still a "certain" (qualitative) agreement, but the discrepancies between experiment and FEA grows significantly. This was traced back to the fact that, after 0.5 ms, the first reflections at the plate boundaries occur, which significantly influence the wave propagation spectrum. Up to the occurrence of first reflections, the boundary conditions of the plate play a minor role, which is not longer the case after reflections.

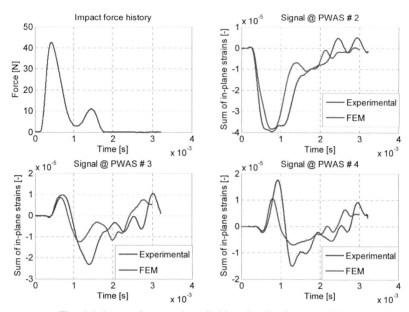

Fig. 8.8: Impact force and strain histories, laminate: "DFG-II-01"

9 Development of a method for the detection of impact events

The residual compressive strength of an impact-damaged composite structure, as shown in Fig. 9.1, has become the design-limiting factor in many cases in the aircraft industry [Rei00]. In contrast aluminum structures, low-velocity impacts on composites often create damages, which are not visible to the naked eye, consisting of internal delaminations and matrix cracks. The detection of such damages requires time-consuming nondestructive testing during which the aircraft has to be out of service. If one would know when impacts occur and their approximate position, then ground inspections could be performed more efficiently.

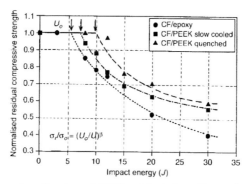

Fig. 9.1: Compression after impact (CAI) residual strength as a function of applied impact   energy [Rei00]

Thus, an impact identification technique for the prediction of the location and force history of the impact would be very helpful in maintaining the structures and reducing inspection costs.

9.1 Arrival time detection

Since an impact is a broad band excitation in the low frequency range, depending on the mass and the contact time of the impactor, flexural waves in a broad band range are excited. Especially in this frequency range, flexural waves are strongly dispersive. Therefore the detection of arrival time of impact induced waves is a challenging task. Most of the techniques referred to in the literature work in the time domain. Gunther et al [Gun93] used the "threshold methods", that is to pick the time when the sensor signal first exceeds an assumed noise level. Schindler et al [Sch95] used the "peak method", that uses the time of the maximum signal as arrival time. Seydel and Chang [Sey01] defined the arrival time by selecting the minimum before the maximum of the signal, and called their approach "double-peak method". However, all the time domain methods have one main disadvantage, namely they cannot be detected to what frequency the detected arrival time belongs. Since flexural waves are strongly dispersive in the frequency range of interest of an impact excitation, the propagation velocity may change significantly as function of frequency even for isotropic materials.

A different approach to define the arrival time is to use the wavelet transform (WT). The theoretical background of WT was given in chapter 2.4.3. Kishimoto *et al* [Kis95] applied the WT for the detection of arrival times; however the approach was restricted to isotropic beams. Gaul and Hurlebaus [Gau98] extended this approach to isotropic plates.

In this research, the approaches developed before are extended to anisotropic composite structures. From each sensor signal, a contour plot using WT is first performed (Fig. 9.2, left). As already mentioned in chapter 2.4.3, the Gabor function was used as mother wavelet. Then a frequency has to be selected at which the magnitude of wavelet coefficients is calculated. Since the higher the frequency the smaller the dispersion, it is advantageous to take a frequency as high as possible in the signal. However, the signal magnitude still has to be strong enough in order to minimize the effect of noise in the signals. Therefore the best compromise has to be found. For the impact hammer excitation performed in this research, the best results were obtained by choosing a frequency of 1.5 kHz. Higher frequencies have much less amplitude and cannot be distinguished clearly from noise, whereas smaller frequencies are much stronger dispersive. Thus, a frequency of 1.5 kHz seemed to be a good compromise. The magnitude of the wavelet coefficients corresponding to that frequency is illustrated in Fig. 9.2 (right).

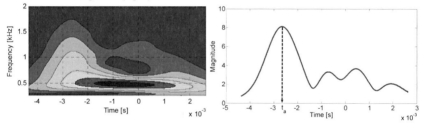

Fig. 9.2: Contour plot of WT (left) and magnitude of wavelet coefficients at selected frequency of 1.5 kHz (right)

The arrival time, t_a, is defined as the first peak of the wavelet coefficients at each sensor.

9.2   Localization of impact events

The next step after the determination of arrival time is the localization of the impacts. For planar impact localization, at least three sensors must be theoretically used. However, in this research four sensors were used in order to minimize the location error, which is also strongly influence by noise and different additional uncertainties. Fig. 9.3 illustrates the geometrical properties of the impacted plate.

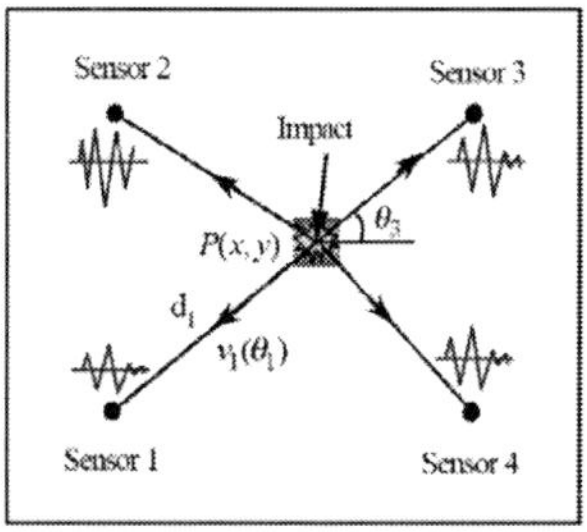

Fig. 9.3: Geometrical plate configuration

The localization of the impact is performed by a triangulation method extended for anisotropic materials, such as composites. Since the velocity of the Lamb waves (predominantly A_0 –mode, respectively flexural waves), induced by the impact, is, additionally to the frequency dependency, a function of propagation direction, the localization is much more difficult than in isotropic materials. With the notations of Fig. 9.3, the following equation can be derived for the theoretical arrival time at each sensor:

$$T_i = \frac{\sqrt{(x - x_i)^2 + (y - y_i)^2}}{v_i(\theta_i)} \qquad (i = 1, 2, 3, 4) \qquad (9.1)$$

Thus, T_i denotes the theoretical arrival time the flexural wave needs to travel from the impact position to the i- th sensor. The velocity $v_i(\theta_i)$ was calculated by the developed higher order plate theory model with the software FIBREWAVE for each 5° propagation direction at different frequencies corresponding to the selected frequencies for arrival time determination. The differential, calculated time of flight, ΔT_i can be then obtained from:

$$\Delta T_i = T_i - T_1 \qquad (i = 2, 3, 4) \qquad (9.2)$$

For the measured differential time of flight, a similar equation can be set up:

$$\Delta t_{ai} = t_{ai} - t_{a1} \qquad (i = 2, 3, 4) \qquad (9.3)$$

where, t_{ai} is the measured arrival time at sensor i. As sensor 1, the sensor is defined with the smallest t_{ai}. The basic idea for the localization of impacts is to minimize the difference between the measured time of flight and the calculated time of flight. The difference which has to be minimized can be expressed by the cost function F as follows:

$$F = \underset{(x,y)}{Min} \sum_{i=2}^{4} (\Delta T_i - \Delta t_{ai})^2 \qquad (9.4)$$

In this research, the minimization problem, was solved by the Nelder-Mead simplex algorithm [Nel65] using MATLAB. Further details on this optimization algorithm are given in chapter 10.1. As initial guessed location for searching the impact, always the x- and y-coordinates of the middle point of the structure were chosen.

9.3 Determination of contact force histories

Besides the estimation of the impact location, it is important to know the force history of an impact. The force history of the impact could be then used for damage prediction, since the delamination area has been shown in many studies [Zho98] to correspond to the maximum impact force and impact energy [Sio98] for low-velocity impacts.

The approach adopted in this thesis is to first derive an analytical structural model, which computes the impact induced strains and then to use it for the reconstruction of impact force histories. As preliminary numerical and experimental studies have already demonstrated, A_0 waves in the low frequency range are almost exclusively excited by a low-velocity impact. In this low frequency range, the accuracy of classical plate theory (CPT) is satisfactory, thus A_0 waves are approximated here as flexural waves. The governing equation is the bending equation for an orthotropic plate, which can be expressed as follows [Pro91]:

$$
D_{11}\frac{\partial^4 w}{\partial x^4} + 4D_{16}\frac{\partial^4 w}{\partial x^3 \partial y} + 2\left(D_{12}+2D_{66}\right)\frac{\partial^4 w}{\partial x^2 \partial y^2}
$$
$$
+ 4D_{26}\frac{\partial^4 w}{\partial x \partial y^3} + D_{22}\frac{\partial^4 w}{\partial y^4} + \rho h \frac{\partial^2 w}{\partial t^2} = q(x,y,t)
$$

(9.5)

where q denotes the distributed load, h the thickness of the plate, ρ the density and w the transverse displacement. This equation is based on classical plate theory which was explained in detail chapter 3.2. If the composite plate is symmetric and has only 0° and 90° plies, then the D_{16} and D_{26} coefficients are both zero. The out-of-plane displacement $w(x,y,t)$ can be then obtained by applying the Rayleigh-Ritz method:

$$
w(x,y,t) = \sum_{i=1}^{I}\sum_{j=1}^{J} A_{ij}(t)\sin\left(\frac{i\pi x}{a}\right)\sin\left(\frac{j\pi y}{b}\right)
$$

(9.6)

where,

$$
A_{ij}(t) = \frac{4}{\rho h a b}\left[\frac{1}{\omega_{ij}}\int_0^t P(\tau)\sin\left[\omega_{ij}\left(t-\tau\right)\right]d\tau\right]\sin\left(\frac{i\pi x_{Im\,pact}}{a}\right)\sin\left(\frac{j\pi y_{Im\,pact}}{b}\right).
$$

(9.7)

For the case when D_{16} and D_{26} are equal to zero, the angular frequency is given by:

$$
\omega_{ij} = \sqrt{\frac{\pi^4}{\rho h}\left[D_{11}\left(\frac{i}{a}\right)^4 + 2\left(D_{12}+2D_{66}\right)\left(\frac{ij}{ab}\right)^2 + D_{22}\left(\frac{j}{b}\right)^4\right]}
$$

(9.8)

Substituting Equation 9.7 into Equation 9.6, one obtains the relation for the transversal displacement of a composite plate at the position x and y at any time t for an impact located at x_{Impact} and y_{Impact}.

$$w(x,y,t) = \frac{4}{\rho h a b} \sum_{i=1}^{I} \sum_{j=1}^{J} \frac{1}{\omega_{ij}} \left[\int_{0}^{t} P(\tau) \sin[\omega_{ij}(t-\tau)] d\tau \right] \sin\left(\frac{i\pi x_{Im\,pact}}{a}\right) \sin\left(\frac{j\pi y_{Im\,pact}}{b}\right) \sin\left(\frac{i\pi x}{a}\right) \sin\left(\frac{j\pi y}{b}\right)$$

(9.9)

The limits for the double summation, I and J, have to be chosen theoretically to infinite, to fully satisfy the equals sign. However, the accuracy is in general sufficient if these coefficients are selected such that they correspond to the highest frequency (see Equation 9.8), which appears in the signal. In this research the coefficient I and J were both set to 10, corresponding to a frequency of about 60 kHz for the used CFRP laminates. In order to include viscoelastic material damping in the analytical structural model, complex material properties were taken into account in Equation 9.8.

However, as already shown in chapter 2.3 of this dissertation, the transverse displacement cannot be measured directly by PWAS transducer, which senses the sum of in-plane strains, i.e. $\varepsilon_x + \varepsilon_y$. Therefore, assuming a pure bending wave excited by the impact, the in-plane strain components can be expressed as:

$$\varepsilon_x(x,y,t) = -z\,\frac{\partial^2 w(x,y,t)}{\partial x^2} \qquad \varepsilon_y(x,y,t) = -z\,\frac{\partial^2 w(x,y,t)}{\partial y^2}$$

(9.10)

After double differentiation of Equation 9.9 with respect to x and y, the following relations are obtained for in plane x- and y- strains:

$$\varepsilon_x(x,y,t) = \frac{4z}{\rho h a b} \sum_{i=1}^{I} \sum_{j=1}^{J} \frac{1}{\omega_{ij}} \left[\int_{0}^{t} P(\tau) \sin[\omega_{ij}(t-\tau)] d\tau \right] \sin\left(\frac{i\pi x_{Im\,pact}}{a}\right) \sin\left(\frac{j\pi y_{Im\,pact}}{b}\right) \sin\left(\frac{i\pi x}{a}\right) \sin\left(\frac{j\pi y}{b}\right)\left(\frac{i\pi}{a}\right)^2$$

(9.11)

$$\varepsilon_y(x,y,t) = \frac{4z}{\rho h a b} \sum_{i=1}^{I} \sum_{j=1}^{J} \frac{1}{\omega_{ij}} \left[\int_{0}^{t} P(\tau) \sin[\omega_{ij}(t-\tau)] d\tau \right] \sin\left(\frac{i\pi x_{Im\,pact}}{a}\right) \sin\left(\frac{j\pi y_{Im\,pact}}{b}\right) \sin\left(\frac{i\pi x}{a}\right) \sin\left(\frac{j\pi y}{b}\right)\left(\frac{j\pi}{b}\right)^2$$

(9.12)

The accuracy of the analytical structural model for the impact loading is studied in the subsection 9.3.1.

In order to determine the impact force history of an impact event based on the signals collected by surface bonded PWAS, an optimization technique was developed using the structural analytical model. Therefore the impact force history is approximated by the following equation:

$$F(t) = F_{max} \cdot \sin\left(\frac{\pi \cdot (t-t_1)}{t_2 - t_1}\right) \text{, where, } t_2 > t_1 \text{ and } t_2 - t_1 = contact\ time$$

(9.13)

Thus, the force history is determined in this model by three parameters: F_{max}, the maximum of the impact force; t_1, the starting point of the impact force (e.g. the time, at which the impactor hits the structure); and t_2, the time at which the impactor "loose" the contact with the structure. The basic idea of the developed impact force detection method is based on non-linear optimization of the parameters F_{max}, t_1 and t_2, which can be expressed as follows:

Find F_{max}, t_1 and t_2 so that the error equation:

$$Error = \sum_{i=1}^{m} \sum_{j=1}^{n} \left| \left[\varepsilon_x\left(x_i, y_i, t_j\right) + \varepsilon_y\left(x_i, y_i, t_j\right) \right]_{PWAS} - \left[\varepsilon_x\left(x_i, y_i, t_j\right) + \varepsilon_y\left(x_i, y_i, t_j\right) \right]_{theor} \right|^2$$

(where n is the number of discrete times measured experimentally and m is the number of PWAS used)
is a minimum.

For optimization, the non-linear Nelder-Mead Simplex optimization method was used. For implementation of the method, MatLab® computational environment was used.

9.3.1   Comparison: Analytical model vs. Experimental results

Since the accuracy of the reconstructed impact force history strongly depends on the physical dynamic model, the accuracy of this model is evaluated first for forward modelling by comparing experimental results collected by PWAS with the analytical model. As force history input, the contact force measured by the instrumented impact hammer is used.

Fig. 9.4 shows the geometry of one of the impact configuration used for testing the analytical method.

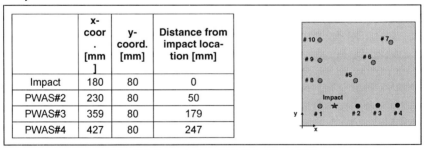

	x-coor. [mm]	y-coord. [mm]	Distance from impact location [mm]	
Impact	180	80	0	
PWAS#2	230	80	50	
PWAS#3	359	80	179	
PWAS#4	427	80	247	

Fig. 9.4: Impact geometrical configuration

The impact test was conducted on the CFRP- laminate with the notation DFG-II-01,(a unidirectional laminate with a thickness of 5.1 mm and 18 unidirectional plies). Using the non-destructive measurements provided by the immersion technique described in chapter 5.2, the complex D_{ij} coefficients were calculated. Since the analytical calculation of impact induced waves is performed by classical plate theory assumptions, only the coefficients shown in Tab. 9.1 are necessary.

D_{11} [kN *mm]	1359.60 + 30.79 i
D_{12} [kN *mm]	40.40 + 8.05 i
D_{66} [kN *mm]	59.69 + 5.53 i
D_{22} [kN *mm]	115.23 + 5.73 i

Tab. 9.1: Structural properties for the specimen tested

As impact force history, the real data measured by the instrumented PCB- impact hammer were used and a numerical integration was performed. According to the measurements conducted experimentally, the sum of the in-plane strains at three PWAS positions (#2, #3, #4) was calculated. At the same time, the voltage output from the PWAS was converted according to the relations highlighted in chapter 2.3 to strains. Fig. 9.5 shows the results obtained experimentally versus the analytical cal- culations, without accounting for viscoelastic material damping. The agreement be- tween analytical solution and measurements is satisfactory; moreover it is obvious that the first peaks for time smaller than 0.4 ms at PWAS #2 and lower than 1 ms at PWAS #3 and #4 agree much better. Since the analytical strain amplitudes were higher than the measured one, additional analytical calculations have been done to account for viscoelastic material damping.

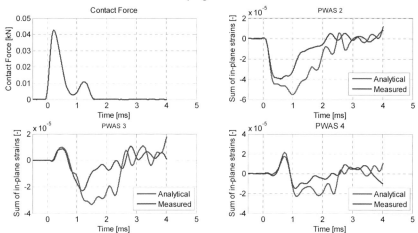

Fig. 9.5: Comparison: Dynamic, analytical model vs. measurements

Fig. 9.6 shows a comparison of measurements with analytical results with and with- out accounting for viscoelastic material damping. The difference is for time smaller than 2.5 ms (for the early stage propagation) negligible. One has to keep in mind, that the wave attenuation of a flexural (or low frequency A_0) wave in the frequency range of such an impact excitation (< 2kHz) is very small (~0.002 dB/mm @ 2 kHz).

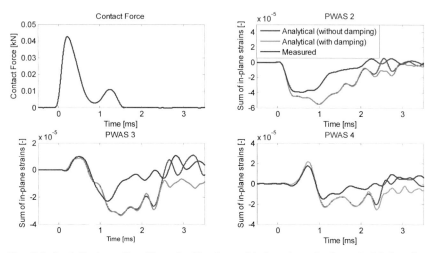

Fig. 9.6: Analytical results with and without considering viscoelastic material damping vs. experimental results

9.4   Testing of the developed method

Every nondestructive evaluation method has to demonstrate a probability of detection of 90% at 95% confidence level in order to be accepted by the aerospace industry [Roa07]. Thus, the goal of this subsection is to test and quantify the performance of the developed impact identification and classification (by impact force reconstruction) approach.

The algorithm and the approach presented in the previous chapter were therefore implemented into a MATLAB software and a vast number of experimental impact tests were performed in order to test and quantify the performance of the developed method.

First, the efficiency of the impact localization method on a CFRP laminate was tested. Basically two types of impacts were given, by the hand held impact hammer and by a small steel ball with a mass of 200 mg. For the impact hammer the arrival time was extracted by using the maximum of the wavelet coefficients corresponding to a frequency of 1.5 kHz, for the steel ball impact, a frequency of 24 kHz was used. Fig. 9.7 shows an example of the localization of a transversal, low-velocity impact given by the impact hammer on a 500 mm x 500 mm CFRP plate.

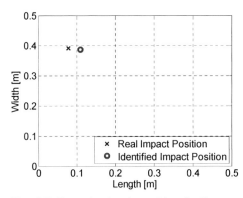

Fig. 9.7: Example of an impact localization result

The impact was applied at x_{real}= 83 mm and y_{real}= 392 mm. The developed and implemented method detected this impact at a position of $x_{detected}$= 105 mm and $y_{detected}$= 387 mm.
In Fig. 9.8 and Fig. 9.9 localization errors for both impact types are shown over many different impacts at several positions.

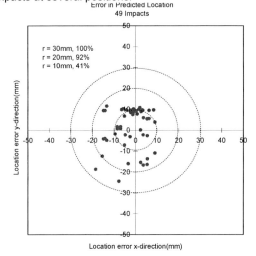

Fig. 9.8: Success rate of the predicted locations over 49 impact tests (steel ball impact)

For the impact with the steel ball, all applied impacts were identified without any exception. All the predicted locations fell within a 30 mm radius from the real impact locations. The predicted locations falling within a 20 mm radius from the real impact locations was about 92%. 41% of the predicted impact fell within a 10 mm radius. The location errors achieved for 49 impacts are illustrated in Fig. 9.8.

Fig. 9.9 shows the location errors determined for 46 impacts given by the instrumented hammer. In this case, all the predicted locations fell within a radius of 50 mm from the actual impact location and 48% fell within a 30 mm radius. However, compared to the steel ball impact, the location errors of the hammer impact are much higher. This was traced back to the following facts:

- the contact time of the hammer impact is longer when compared to the steel ball impact. The maximum of the impact force arrives at the PWAS together with reflections from the boundaries and therefore the arrival time detection is much more difficult.
- the frequencies excited by hammer impacts are much lower (< 3 KHz) in comparison to those excited by the steel ball impacts (<40 kHz). Thus, for the arrival time detection by wavelet coefficients, a lower frequency of 1.5 kHz had to be chosen at which dispersion of the flexural waves is much stronger. For the steel ball impact a much higher frequency of 24 kHz could be adopted for which dispersion of flexural waves, respectively A_0 Lamb wave mode is much less.

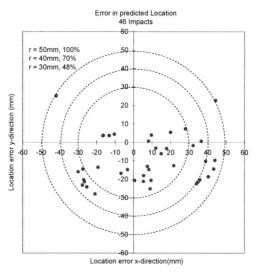

Fig. 9.9: Success rate of the predicted locations over 46 impact tests (instrumented hammer impact)

After successful verification and quantification of the impact location abilities of the developed method, experimental tests of the impact force reconstruction method were conducted. Since for comparison purposes the actual applied impact force was needed, all impacts were applied by the instrumented impact hammer. Fig. 9.10 and Fig. 9.11 show the identified vs. applied impact force, demonstrating a satisfactory agreement.

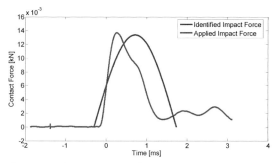

Fig. 9.10: History of identified and applied impact force

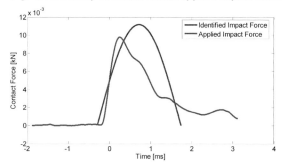

Fig. 9.11: History of identified and applied impact force

10 Further applications of Lamb waves: In-situ quantitative viscoelastic material characterization

The objective of the previous chapters of this research was to calculate dispersion and attenuation curves for composites assuming that the viscoelastic material properties are known. In this context it was observed that the dispersion curves (velocity and attenuation) are strongly affected by the viscoelastic properties of the material of the specimen. In this chapter, the inverse problem, i.e. an approach to determine the viscoelastic material properties from experimentally obtained dispersion curves will be discussed.

State of the art conventional techniques for the determination of material properties are costly, destructive and often inaccurate, which is particularly true for the detection of through the thickness properties of composites. For this purpose a nondestructive quantitative material characterization technique may offer many advantages. It is repeatable and appropriate for in-service/in-situ use (e.g. as a component of a SHM-system), allowing to determine the degradation of the stiffness parameters of the structure with respect to time. Such degradation or damage factors are moreover essential for the design of a structure and in explicit finite element methods.

The use of guided waves for material property characterization in thin samples was first demonstrated by Kubiak, Hosek and Lichodzielewski [Kub66] in 1966. They constructed a frequency modulated inspection system and showed excellent sensitivity to defects, thickness variations, and anisotropic material properties. In more recent years, Wang and Rokhlin [Wan90] presented a method for the elastic characterization of anisotropic layers. They applied their method to a porous aluminum oxide layer 50 µm in thickness. Kinra and Dayal [Kin90] presented a method for determining the dynamic elastic moduli of a number of materials through body-wave-based ultrasonic experiments. Another often reported technique is the so-called leaky Lamb wave technique (LLW), which is based on the measurement and analysis of guided waves in thin specimens [Nay88], [Pil88], [Mal89], [Kar90]. The LLW technique is based on an oblique insonification of a test specimen that is fully immersed in water. However, the only work presented for viscoelastic material characterization was presented by Castaings and Hosten [Cas00], [Cas01] at the University of Bordeaux. Their method is based on the immersion technique and is the method adopted for the initial material characterization in this dissertation. However, this method is not appropriate for in-situ material characterization. A comprehensive summary of the use of guided waves for materials characterization is given in the review article by Chimenti [Chi97].

The intention of this chapter is to analyze a novel material characterization method based on in-situ experimentally obtained velocity and attenuation dispersion data. The dispersion data are obtained with surface bounded PWAS, as described in chapter 6. The theoretical relationship between material properties (i.e. viscoelastic stiffness constants C^*_{ij}) and dispersion curves has been provided by the developed higher-order plate theory including material damping and implemented in the software FIBREWAVE. The equations provided in this manner are inverted by means of an optimization scheme which minimizes the error between theoretical and experimental dispersion curves.

The minimization problem can be expressed as follows:

$$F(C_{ij}) = Min ! \sum_{i=1}^{n} \left| c_g(i)_{(Experimental)} - c_g(i)_{(Theoretical)} \right|^2 \tag{10.1}$$

where n is the total number of wave frequencies, for which experimental data are available and for which theoretical values have to be provided.

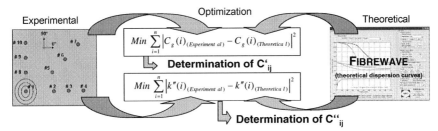

Fig. 10.1: Proposed of a quantitative characterization method for viscoelastic composites

10.1 Mathematical method for function minimization

For the inversion of the experimental Lamb wave data, appropriate optimization techniques are necessary. For this purpose, several techniques are available in the literature, such as the steepest descent and Newton-Raphson methods, or a combination of the two methods, such as the Marquardt algorithm [Mar63] were reviewed. Most of the nonlinear inversion techniques have the main disadvantage that the unknown parameters must be adjusted in an iterative way with little advance knowledge of how many repetitions will be needed for convergence, or whether convergence would be achieved at all. An exception is the Nelder-Mead simplex algorithm (NMSA), first published in 1965 by John Nelder and Roger Mead [Nel65], which is intended for multidimensional, unconstrained function minimization as required in this research for the inversion of Lamb wave data for obtaining (viscoelastic) stiffness coefficients. The NMSA minimizes a scalar-valued nonlinear function of n real variables using only function values, thus without any derivative information (derivative-free method). This property is crucial for the inversion of Lamb wave data, since the developed wave propagation model cannot be formulated in a closed form, hence there is no analytical function available to differentiate. The NMSA falls in the general class of direct search methods.

Since it is neither the intention of this research to develop a novel mathematical optimization algorithm nor to give a comprehensive review of the NMSA, the reader is referred for further information to [Nel65] and to [Lag98]; the latter gives a contemporary view of the algorithm and discusses its convergence properties in detail.

In this research, the Nelder-Mead simplex algorithm was applied using MATLAB's *fminsearch*, which is part of the Optimization Toolbox.

10.2 Sensitivity analysis

In order to determine accurately the material properties using the proposed method, it is essential first to ensure that the dispersion curves are sensitive enough to the viscoelastic coefficients. It must be also determined which Lamb wave modes and propagation angles are most sensitive to each specific coefficient. Therefore a sensitivity analysis with the software FIBREWAVE has been performed for the unidirectional laminate "DFG-II-02". During this analysis, each stiffness coefficients was separately reduced by 50% of its original value, showing in this way its influence on the dispersion curves. Fig. 10.2 illustrates the influence of the real part of the stiffness constants C'_{ij} on the A_0, SH_0 and S_0 (velocity) dispersion curves at 0° propagation angle (i.e. propagation parallel to the fibers).

As the parameter study in Fig. 10.2 shows, some of the stiffness constants (e.g. C'_{11}, C'_{13}, C'_{23}, C'_{55}, C'_{66}) are very sensitive to specific dispersion curves and relatively insensitive to other wave modes, whereas other constants (e.g. C'_{12}, C'_{22}, C'_{33}, C'_{44}) are insensitive to all three considered dispersion curves at wave propagation parallel to the fibers. Since in anisotropic laminates, such as unidirectional CFRP laminates, the dispersion curves are a function of propagation angle, the same sensitivity parameter study was performed for 90° propagation angle, e.g. perpendicular to the fibers. The parameter sensitivity study at 90° propagation angle is shown in Fig. 10.3.

In principle, a similar behavior is observed. Some of the stiffness coefficients are strongly involved in some dispersion curves, some other coefficients are relatively insensitive, like C_{11}, C_{12}, C_{13} and C_{55}. Of course the insensitive constants would not be able to be determined.

Additionally to the sensitivity study shown in Fig. 10.2 and Fig. 10.3 the sensitivity to the dispersion curves at 45° propagation angle was investigated. The corresponding figures can be found in the Appendix.

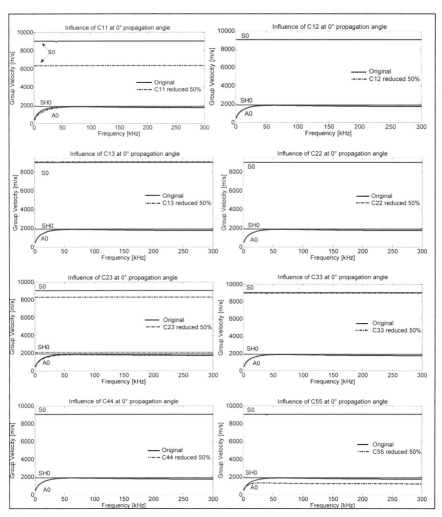

Fig. 10.2: Influence of stiffness coefficients on dispersion curves at 0° propagation angle

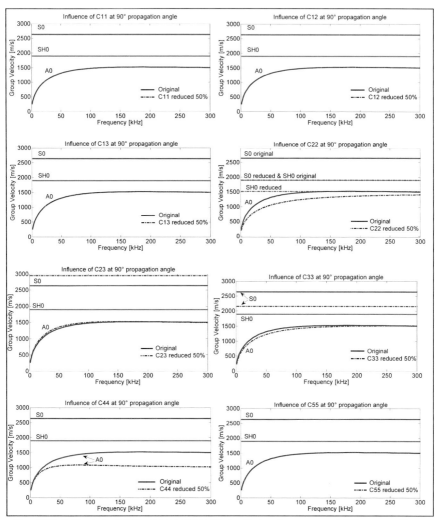

**Fig. 10.3: Influence of stiffness coefficients on dispersion curves at 90° propagation
angle**

In Tab. 10.1 a summary of the influences of all stiffness coefficients classified as
"strong influence", "some/very less influence" and "no influence at all" is given.

Influence	0° propagation Angle ($\parallel$ to the fibers)			90° propagation angle ($\perp$ to the fibers)			45° propagation angle to the fibers			Legend
	S_0	SH_0	A_0	S_0	SH_0	A_0	S_0	SH_0	A_0	
C'_{11}	xx	o	x	o	o	o	x	x	x	xx : strong
C'_{12}	o	o	o	o	o	o	o	o	o	Influence
C'_{13}	x	o	o	o	o	o	x	o	o	
C'_{22}	o	o	o	xx	x	x	o	xx	o	x : some/very
C'_{23}	xx	x	o	xx	o	x	o	x	o	less influence
C'_{33}	o	o	o	xx	o	x	x	x	o	
C'_{44}	o	o	o	o	o	xx	o	o	x	o : no influ-
C'_{55}	o	o	xx	o	o	o	o	o	xx	ence at all
C'_{66}	o	x	o	o	x	o	o	x	o	

Tab. 10.1: Influences of C'_{ij} on S_0, SH_0 and A_0 velocity dispersion curves of a unidirectional CFRP laminate

Thus, all but C'_{12} were found to have an influence on the S_0, SH_0 or A_0 dispersion
curves. According to Tab. 10.1, the stiffness constants C'_{11}, C'_{23}, $C'_{22,}$ C'_{33} may be
accurately determined through inversion of S_0 dispersion curves, since these dispersion curves are sensitive to these stiffness coefficients and relatively insensitive to
the other ones. From inversion of the A_0 dispersion curves the stiffness coefficients
C'_{44} and C'_{55} might be obtained and from SH_0 mode the C'_{22} coefficient could be obtained. For the stiffness coefficients C'_{13} and C'_{66} only some or very little influence on
any dispersion curves was observed. Since the influence of these coefficients is supposed to be within the standard deviation obtained by experiments, one can assume
that their values cannot be determined accurately.

Analog to the sensitivity parameter study for the real part of the stiffness coefficients
(C'_{ij}), the influence of the imaginary parts (C''_{ij}) on the attenuation dispersion curves
was investigated. The results are summarized in Tab. 10.2.

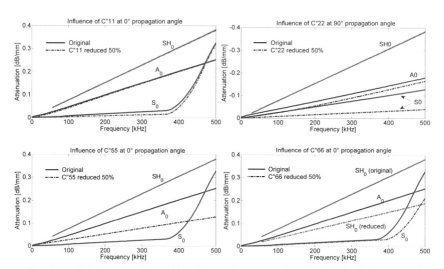

Fig. 10.4: Selection of attenuation dispersion curves illustrating the influences of
imaginary part stiffness coefficients

	0° propagation Angle (∥ to the fibers)			90° propagation angle (⊥ to the fibers)			45° propagation angle to the fibers			Legend
Influence	S_0	SH_0	A_0	S_0	SH_0	A_0	S_0	SH_0	A_0	
C''_{11}	xx	o	o	o	o	o	xx	o	o	xx : strong
C''_{12}	o	o	o	o	o	o	o	x	o	Influence
C''_{13}	o	o	o	o	o	o	o	o	o	
C''_{22}	o	o	o	xx	o	x	o	xx	o	x : some/very
C''_{23}	o	o	o	xx	o	x	o	xx	x	less influence
C''_{33}	o	o	o	xx	o	x	o	xx	o	
C''_{44}	o	o	o	o	o	xx	o	o	x	o : no influ-
C''_{55}	o	o	xx	o	o	o	o	o	xx	ence at all
C''_{66}	o	xx	o	o	xx	o	o	xx	x	

Tab. 10.2: Influences of C''_{ij} on S_0, SH_0 and A_0 attenuation dispersion curves of an
unidirectional CFRP laminate

As shown in Tab. 10.2 some of the C''_{ij} coefficients possess strong influences on
more than one Lamb wave type and for more than one wave propagation, as for example C''_{11}, which has a strong influence on the S_0 mode in 0° propagation direction
as well as in 45° direction. However, when more than one C''_{ij} coefficient has strong
influence on a certain dispersion relation, such as C''_{22}, C''_{23}, and C''_{33} on the S_0
mode in 90° direction, then the solution may be non-unique, e.g. different combina-
tions of C''_{ij} coefficients may give the same results.

10.3 Infer Viscoelastic material properties

This research deals only with the inversion of viscoelastic material properties of unidirectional laminates. Moreover, in order to decrease the degrees of freedom in the
optimization procedure, it is assumed that the unidirectional laminates are transversely isotropic. This assumption allows to reduce the number of stiffness coefficients to describe the viscoelastic material behavior from nine coefficients to five
(C_{11}, C_{12}, C_{23}, C_{22}, C_{55}). The remaining four coefficients can be calculated as follows:

$$C_{13} = C_{12}, \qquad C_{33} = C_{22}, \qquad C_{66} = C_{55}, \qquad C_{44} = \frac{C_{33} - C_{23}}{2} \qquad (10.2)$$

The concept for the inversion of viscoelastic coefficients developed in this research is
to perform the inversion in two main steps. The first step is to obtain the stiffness coefficients, i.e. the real parts of C_{ij} and the second step is to invert the experimental
attenuation data in order to obtain the attenuation coefficients, i.e. the imaginary part
of C_{ij}.

10.3.1 Concept for obtaining the stiffness coefficients (i.e. real part of C_{ij})

As mentioned before, the first inversion step is to invert the (velocity) dispersion data
in order to obtain the real part of the C_{ij} coefficients. As shown in Tab. 10.1, all but
C'_{12} have a strong influence on at least one dispersion relation. Hence, the C'_{12} coefficient cannot be obtained by this technique, but for all four remaining coefficients
there is a strong influence on at least one velocity dispersion curve. According to the
influences and dependencies shown in Tab. 10.1, the inversion of C'_{ij} coefficients
was performed in the following order:
1. C'_{11} from S_0 mode in 0° with CLT
2. C'_{22} from S_0 mode in 90° with CLT
3. C'_{55} from A_0 mode in 0° with HPT/FibreWave
4. C'_{44} from A_0 mode in 90° with HPT/FibreWave
5. C'_{23} from transversely isotropy relation: $C'_{23} = C'_{22} - 2 * C'_{44}$
6. go back to step 1. and use obtained values as new starting/guessed values, if
   accuracy > limit

However, as shown above, the inversion of C'_{11} and C'_{22}, was performed using the
simple classical plate theory (CLT), since in the low-frequency range (< 150 – 200
kHz), the CLT provides sufficient accuracy and a closed form solution. The flowchart
of the inversion approach for obtaining the real part of C_{ij} is illustrated in Fig. 10.5.

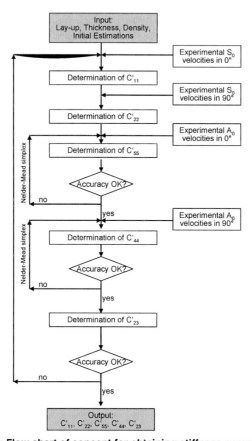

Fig. 10.5: Flow chart of concept for obtaining stiffness parameters C'$_{ij}$

For the inversion of C'$_{11}$ and C'$_{22}$ there is no optimization procedure necessary and no initial estimation, since the CLT provides a closed form solution and possess no frequency dependency. Therefore the mean values of the experimentally obtained data in the frequency range <200 kHz was used. For obtaining the coefficients C'$_{55}$ and C'$_{44}$, the Nelder-Mead simplex algorithm was used as pointed out in chapter 10.1. The C'$_{23}$ coefficient was determined from coefficients C'$_{33}$ (resp. C'$_{22}$) and C'$_{44}$ using the relation for transversal isotropy. After the first global iteration is finished, additional iterations are performed until the coefficients do not change any more (in a certain limit). Tab. 10.3 shows the inversion of the real part of the C$_{ij}$ coefficients using the experimental data of the CFRP- laminate "DFG-II-02" with 14 iterations. The initial estimation used for the C'$_{55}$ and C'$_{44}$ coefficients was 6 GPa for both.

Iteration no.	C'$_{11}$ [GPa]	C'$_{22}$ [GPa]	C'$_{55}$ [GPa]	C'$_{44}$ [GPa]	C'$_{23}$ [GPa]
0 (Initial estimation)	No initial Estimation	No initial estimation	6.00	6.00	No initial estimation
1	127.32	12.95	5.94	3.44	6.07
2	127.31	13.01	5.93	3.43	6.14
4	127.29	13.09	5.93	3.44	6.21
6	127.28	13.14	5.93	3.44	6.27
8	127.27	13.17	5.93	3.44	6.30
10	127.27	13.19	5.93	3.44	6.32
12	127.26	13.20	5.93	3.44	6.33
14	127.26	13.21	5.93	3.44	6.33

Tab. 10.3: Optimization for obtaining stiffness constants

As illustrated in Tab. 10.3, the C'$_{11}$ coefficient shows a change of only -0.047% by comparing the result of the 1st iteration with that after the 14th iteration. However, the C'$_{22}$ coefficient changed during the same number of iteration by 2.01%. A similar behavior show the C'$_{55}$ and C'$_{44}$ coefficients compared to the C'$_{23}$. The reason for this behavior is that the higher the influence of other coefficients on the corresponding dispersion relation the more may change the coefficient to be optimized. Fig. 10.6 show the theoretical group velocity data after 14 iterations compared to the experimental data.

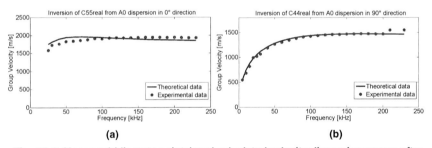

(a)   (b)

Fig. 10.6: Measured (discrete points) and calculated velocity dispersion curves after 14 iterations:
(a) A$_0$ (0°-fibre direction) dispersion data for obtaining C'$_{55}$ coefficient;
(b) A$_0$ (90°-matrix direction, in-plane) dispersion data for obtaining C'$_{44}$ coefficient

10.3.2 Concept for obtaining the damping coefficients (i.e. imaginary part of C$_{ij}$)

The concept for obtaining the damping coefficients (i.e. the imaginary parts of C$_{ij}$) is very similar to the one for the inversion of stiffness coefficients, presented in chapter 10.3.1. The main difference is that instead of using experimental velocity dispersion data, experimental attenuation dispersion data have to be used in order to obtain the imaginary parts of the C$_{ij}$ coefficients.

After the sensitivity analysis performed in chapter 10.2 and summarized in Tab. 10.2, the following concept was developed to obtain the C''_{ij} coefficients:

1. C''_{11} from S_0 mode in 0° direction
2. C''_{55} from A_0 mode in 0° direction
3. C''_{22} and C''_{23} from S_0 mode in 90° direction
4. go back to step 1. and use obtained values as new starting/guessed values, if accuracy > limit

Preliminary analysis has shown that all the remaining C''_{ij} coefficients, which are not mentioned above, cannot be determined by this method, due to lack of sensitivity. In Fig. 10.7 a flowchart of the optimization concept is provided.

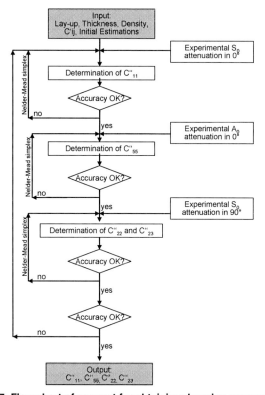

Fig. 10.7: Flow chart of concept for obtaining damping parameters C''_{ij}

The Nelder-Mead simplex algorithm is used for the "internal" optimization loop for the determination of single coefficients using experimental attenuation data from a single Lamb wave mode in a single direction. The "external" loop aims to start again the whole optimization process using the C''_{ij} coefficients from previous optimization steps as initial values. Tab. 10.4 shows the results for the DFG-II-02 laminate using

for all imaginary parts of C_{ij} the same initial estimation of 0.50 GPa after the first iteration step.

Iteration no.	C''_{11} [GPa]	C''_{55} [GPa]	C''_{22} [GPa]	C''_{23} [GPa]
Initial estimation	0.50	0.50	0.50	0.50
1	3.4750	0.1094	0.5304	0.4748

Tab. 10.4: Optimization for obtaining damping constants

Fig. 10.8 illustrates the theoretical attenuation data vs. the experimental ones, which
have been used for the determination of the C''_{ij} coefficients, after the final iteration.

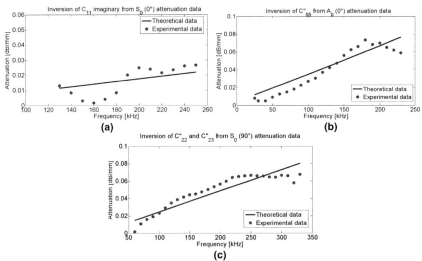

Fig. 10.8: Measured (discrete points) and calculated attenuation dispersion curves
after 14 iterations:
(a) S_0 (0°-fibre direction) attenuation dispersion data for obtaining C''_{11};
(b) A_0 (0°-fibre direction) attenuation dispersion data for obtaining C''_{55};
(c) S_0 (90°-matrix direction, in-plane) attenuation dispersion data for obtaining C''_{22} and
C''_{23}

10.4 Comparison of obtained results with other methods

The intention of this chapter is to compare the C_{ij} values obtained by the nondestructive inversion technique developed in this research with the results from other methods. For the measurement of stiffness coefficients (only 2D) simple destructive tensile tests in 0°, 45° and 90° direction have been performed. Of course, it is not possible to obtain the damping coefficients by this (quasi-static) method. For obtaining the
damping coefficients as well as the stiffness coefficients the non-destructive immersion technique was adopted. This method was presented in chapter 5.2.

Tab. 10.5 shows the comparison between the results obtained by the optimization
method presented in this research and denoted "Optimization using PWAS", and the
"immersion technique". As already mentioned, due to lack of sensitivity of the guided

wave modes A_0, S_0 and SH_0 on the C'_{12} and C'_{13} stiffness coefficients, these values cannot be obtained by this method. Hence, these coefficients have to be provided by additional mechanical destructive or non-destructive tests.

	Optimization using PWAS	Immersion technique	Difference [%]	Within tolerance of immersion technique?
C'_{11} [GPa]	127.26	125.0 $^{\pm 8}$	1.81 %	YES
C'_{22} [GPa]	13.21	13.9 $^{\pm 0.4}$	-4.96 %	NO
C'_{33} [GPa]	13.21	14.5 $^{\pm 0.4}$	-8.90 %	NO
C'_{12} [GPa]	cannot be determined	6.3 $^{\pm 0.6}$	-	-
C'_{13} [GPa]	cannot be determined	5.4 $^{\pm 0.4}$	-	-
C'_{23} [GPa]	6.33	7.1 $^{\pm 0.2}$	-10.85 %	NO
C'_{44} [GPa]	3.44	3.7 $^{\pm 0.3}$	-7.03 %	YES
C'_{55} [GPa]	5.93	5.4 $^{\pm 0.4}$	9.81 %	NO
C'_{66} [GPa]	5.93	5.4 $^{\pm 0.6}$	9.81 %	YES
C''_{11} [GPa]	3.4750	3.0 $^{\pm 1}$	15.83 %	YES
C''_{22} [GPa]	0.5304	0.6 $^{\pm 0.2}$	-11.60 %	YES
C''_{33} [GPa]	0.5304	0.6 $^{\pm 0.2}$	-11.60 %	YES
C''_{12} [GPa]	cannot be determined	0.9 $^{\pm 0.6}$	-	-
C''_{13} [GPa]	cannot be determined	0.4 $^{\pm 0.2}$	-	-
C''_{23} [GPa]	0.4748	0.23 $^{\pm 0.15}$	106.43 %	NO
C''_{44} [GPa]	0.0278	0.12 $^{\pm 0.06}$	-76.83 %	NO
C''_{55} [GPa]	0.1094	0.3 $^{\pm 0.15}$	-63.53 %	NO
C''_{66} [GPa]	0.1094	0.5 $^{\pm 0.25}$	-78.12 %	NO

Tab. 10.5: Comparison "Optimization using PWAS" vs. "Immersion technique" for the CFRP laminate "DFG-II-02" with $[0°]_{12}$

The agreement of the stiffness coefficients to the values obtained by the immersion technique is excellent, with an error tolerance below 10%. In particular, the C'_{11} coefficient can be determined much more accurate than by the immersion technique. As expected, the imaginary part of C_{ij} is a much more challenging task, which is due to the fact that, experimentally, it is much more difficult to obtain reliable attenuation measurements than velocity measurements. The wave velocity can be calculated from the arriving time of the wave, whereas the attenuation requires an accurate measurement of the wave amplitude. The accuracy of the latter task is much more influenced by noise and other uncertainties, than the first one. However, for the determination of the imaginary parts, the immersion technique also gives much higher error, as shown in Tab. 10.5 (in particular for the C''_{23} value with the tolerances C''_{23} = 0.23 $^{\pm 0.15}$).

The third method adopted in this research work for the determination and comparison of stiffness coefficients is the conventional, quasi-static tensile test, Tab. 10.6. The main advantage of the tensile test is that it provides not only stiffness, but also strength properties of the material. Of course, one disadvantage is that the method is destructive.

	Optimization using PWAS	Immersion technique	Tensile Test (Destructive)
E'_{11} [GPa]	123.26	121.74	122.15
E'_{22} [Gpa]	10.09	10.32	9.37
E'_{33} [Gpa]	10.09	10.84	-
G'_{12} [Gpa]	5.93	5.4	5.7
G'_{13} [Gpa]	5.93	5.4	-
G'_{23} [Gpa]	3.44	3.7	-
v'_{12}	0.32242	0.35325	0.2800
v'_{21}	0.026406	0.029837	0.0253
v'_{31}	0.026406	0.017822	-
v'_{32}	0.46659	0.50242	-
v'_{13}	0.32242	0.19978	-
v'_{23}	0.46659	0.47841	-

Tab. 10.6: Comparison of engineering constants obtained by different techniques on the CFRP laminate "DFG-II-02" with lay-up $[0°]_{12}$

Finally it can be stated, that the optimization method using PWAS transducers, besides some uncertainties and limitations, could be used for the in-situ, non-destructive and quantitative characterization of unidirectional fiber reinforced laminates.

11 Conclusions

Structural Health Monitoring (SHM) has become an important issue in recent decades, particularly in the aerospace industry and civil engineering. Various SHM systems have already been developed for numerous potential applications. Many prototypes and laboratory demonstrations have been presented and displayed at meetings, conferences and workshops. Despite all these significant recent advances, there is still a need of fundamental understanding and of developing new applications, which was the motivation of this research.

The focus of this dissertation lies on the fundamental understanding of viscoelastic Lamb waves in anisotropic composites. In particular, the study of the attenuation of such guided waves was the main objective of this research. As an example for the application of Lamb Waves for SHM issues the following two methods were developed, implemented and demonstrated:
- a method for identification and classification of impacts
- a novel method for non-destructive anisotropic material characterization.

11.1 Summary of performed work

Since one of the priorities of this work lied in the analytical description of Lamb waves propagation in anisotropic viscoelastic composite materials, a higher order plate theory was developed taking into account transversal shear and rotational inertia. Material damping was considered through the theory of complex moduli. For the prediction of dispersion and attenuation behavior of Lamb waves, the developed models were implemented into a software enabling the calculation of S_0, A_0 and SH_0 modes. The results were verified experimentally showing a good agreement. For experimental verification, an experimental set-up was built up, in which Lamb waves could be excited and their propagation was measured by small and lightweight, surface-bonded piezoelectric waver active sensors (PWAS). The frequency ranged from 10 – 500 kHz. Moreover, the dispersion and attenuation behavior of Lamb wave propagation was experimentally and analytically studied for six carbon fiber reinforced plastic (CFRP) laminates, which are typical in the aerospace industry.

In order to detect and quantify impact events, an existing method was further developed for application on anisotropic, viscoelastic composites. The flexural waves excited by the impact were measured by surface bonded PWAS and the signals were analyzed by the Wavelet transform (WT). For the reconstruction of the impact force history an analytical structural model, including material damping, was adopted. A verification of the proposed methodology showed good agreement with the real impact position and force time history.

Another important issue in this dissertation was the development of a novel non-destructive, quantitative viscoelastic material characterization method based on Lamb wave dispersion and attenuation. Basically, the method represents the inverse solution of the analytical, higher order plate model, as developed before. The inversion of the viscoelastic material stiffness coefficients was realized by a non-linear simplex optimization method. The results were applied to an unidirectional CFRP laminate, since the approach was limited to unidirectional laminates. The developed method has the potential to be applied in-situ and become one of the components of future SHM systems.

The major findings in this dissertation can be summarized as following:

- a hysteretic damping material model formulated with complex moduli can be used for accounting of material damping of composites
- a higher order plate theory (HPT) can be applied to compute A_0, S_0 and SH_0 Lamb wave modes in viscoelastic composites
- the HPT offers some advantages in comparison to the "exact" 3-D theory and semi-analytical methods, such as semi analytical finite elements (SAFE) methods in term of computational costs without reducing significantly the accuracy of the results
- for all CFRP laminates studied, the S_0 mode shows the smallest attenuation, whereas the attenuation levels of SH_0 and A_0 strongly depend on the lay-up of the laminate
- the application of the Wavelet transform increases significantly the accuracy for arrival time detection for impact localization applications on composites
- the inverse problem of the Lamb wave computational model using HPT and a hysteretic material damping model can be solved almost in real time for inverting viscoelastic stiffness data for given Lamb wave dispersion and attenuation curves. The Mead –Nelder simplex algorithm was used successfully for the inversion.

11.2 Contributions to the advancement of the state of the art

The main contributions of this dissertation to the advancement of the state of the art are listed as following:

- Development and implementation of a higher order plate theory for viscoelastic Lamb wave propagation in anisotropic materials.
- Experimental demonstration of the applicability of the higher order plate theory for the prediction of anisotropic Lamb wave attenuation in composites.
- Extensive experimental Lamb wave attenuation measurements on different CFRP laminates with PWAS transducers.
- Development of a quantitative inverse-solution for a material characterization method in order to obtain viscoelastic material properties from experimental Lamb wave velocity and attenuation dispersion curves. This method can be applied in-situ.
- Development of a method for impact detection, location determination and force reconstruction in composite materials.

11.3 Suggested further work and applications

As usual in every research, there are still questions to be resolved and further work must be done to fully discover the potential and further applications of Lamb wave based SHM in composites. The models developed in this research are limited to a single attenuation mechanism: the attenuation due to material damping. However, attenuation of guided waves occurs by different other mechanisms, which have been studied very little yet. The attenuation of Lamb wave due to energy losses into adjacent media (e.g. losses in pipes due to the contained fluid or structural attenuation at stiffeners and joints) is for example one of the mechanisms very little studied. Moreover, the quantitative increase of Lamb wave attenuation in composites due to defects is not sufficiently studied yet.

The models developed and implemented in this research could be implemented and become part of an approach for sensor network optimization, in which the number and location of sensors for a required probability of detection (POD) is computed. The application of the viscoelastic higher order plate theory on higher order guided wave modes, such as S_1, A_1, SH_1, S_2... has to be evaluated.

As mentioned before, the developed material characterization method was limited to unidirectional laminates. By using additional Lamb wave dispersion and attenuation data, it could be extended to laminates with arbitrary lay-ups. Moreover the method can be applied to determine the material stiffness and attenuation properties over time, in order to evaluate aging effects. The obtained results can be used in life-time prediction models and finite element analysis.

REFERENCES

[Aal73]   Aalami, B.: Waves in prismatic guides of arbitrary cross section, *Journal of Applied Mechanics*, Vol. 40, pp. 1067-1072, 1973.

[Ach99]   Achenbach, J. D.: *Wave Propagations in Solids*, North-Holland, 1999.

[All92]   Alleyne, D. N., Cawley, P.: Optimization of Lamb Wave inspection techniques, *NDT&E International*, Vol. 25, No. 1, pp. 11-22, 1992.

[All92a]   Alleyne, D. N., Cawley, P.: The Interaction of Lamb Waves with Defects, *IEEE Transactions on Ultrasonics, Ferroelectrics and Frequency Control*, Vol. 39, No. 3, pp. 381-396, 1992.

[Apc07]   www.americanpiezo.com/materials/apc_properties.html, 2007.

[Ash70]   Ashton, J. E., Whitney, J. M.: *Theory of Laminated Plates,* Stamford, Conneticut, Technomic Publishing Co., 1970.

[Bal06]   Balageas, D., Fritzen, C. P., Güemes, A.: *Structural Health Monitoring*, ISTE Ltd., ISBN 978-1-905209-01-9, 2006.

[Bar06]   Bartoli, I., Marzani, A., Lanza di Scalea, F., Viola, E.: Modeling wave propagatio in damped waveguides of arbitrary cross-section, *Journal of Sound and Vibration*, Vol. 295, pp. 685-707, 2006.

[Boc06]   Bockenheimer, C.: Airbus Needs from Nanotechnology for Structural Health Monitoring (SHM), *Séminaire "Nanotechnologies en France",* Nanotech, 2006.

[Bot05]   Bottai, G. S., Giurgiutiu, V.: Lamb wave interaction between piezoelectric wafer active sensors and host structure in a composite material, *Proc. of 5th International Workshop on Structural Health Monitoring*, Stanford, September 2005.

[Cal05]   Calomfirescu, M., König, C., Müller, J.: Entwicklung von Gestaltungsrichtlinien für impacttolerante Faserverbundstrukturen [in German], *Proc. of 8th International AVK-TV Conference*, Baden-Baden, 2005.

[Cal06]   Calomfirescu, M., Herrmann, A. S., König, C., Müller, J.: Development of a FEA-based approach to localize impacts and determine their force histories, *Proc. of Adaptronic Congress*, Göttingen, May, 3-4, 2006.

[Cal06a]   Calomfirescu, M., Herrmann, A. S., König, C., Müller, J.: Numerische und experimentelle Untersuchung der Wellenausbreitung in Faserverbundstrukturen, [in German], *Materialprüfung*, Vol. 5, May 2006.

[Cal06b]   Calomfirescu, M., Herrmann, A. S., König, C., Müller, J.: Investigation on Wave Propagation in Composites as Requirement for Impact detection, *Proc. of 25th International Congress of the Aeronautical Sciences*, Hamburg, September 2006.

[Cal07]   Calomfirescu, M., Herrmann, A.S.: Theoretical and experimental studies of Lamb wave propagation in attenuative composites, *Proc. of SPIE*, San Diego, March 2007.

[Cal07a]   Calomfirescu, M., Herrmann, A.S.: Propagation of Lamb Waves in viscoelastic Composites for Impact Monitoring Applications, *Prof. of SAMPE Europe*, Paris, March 2007.

[Cal07b]   Calomfirescu, M., Herrmann, A. S.: On the propagation of Lamb Waves in viscoelastic composites for SHM applications, *Key Engineering Materials*, Vol. 347, pp. 543-548, 2007.

[Cal07c]   Calomfirescu, M., Herrmann, A. S.: Attenuation of Lamb Waves in Composites: Models and possible Applications, *Proc. of 6th International Workshop on Structural Health Monitoring*, Stanford, September 2007.

[Cal08]   Calomfirescu, M.: Characterization of viscoelastic Unidirectional Composite Laminates using S_0, A_0 and SH_0 Lamb waves, *Proc. of 4th European Workshop on Structural Health Monitoring*, Cracow, July 2008 [in press].

[Cas00]   Castaings, M., Hosten, B., Kundu, T.: Inversion of ultrasonic, plane-wave transmission data in composite plates to infer viscoelastic material properties, *NDT & E International*, Vol. 33(6), pp. 377-392, 2000.

[Cas01]   Castaings, M., Hosten, B.: Lamb and SH waves generated and detected by air-coupled ultrasonic transducers in composite material plates, *NDT&E International*, Vol. 34, pp, 49-258, 2001.

[Cas07]   Castaings, M., Le Crom, B.: *Characterisation of viscoelastic properties of composite materials*, Internal Report, Ordering nbr UM/Faserinstitut Bremen, January 2007.

[Cha02]   Chang, F. K.: Ultra reliable and super safe structures for the new century, *Proc. of the First European Workshop on Structural Health Monitoring*, Cachan, France, pp. 3-12, 2002.

[Cha95]   Chang, F. K.: Built-in damage diagnostics for composite structures, *Proc. of the 10th Int. Conf. on Composite Structures (ICCM-10)*, Vol. 5, Canada, 14-18 August, 1995.

[Cha98]   Chang, F. K.: Manufacturing and design of built-in diagnostics for composite structures, *Proc. of the 52nd Meeting of the Society for Machinery Failure Prevention Technology*, Virginia Beach, VA, 30 March – 3 April, 1998.

[Cha98a]   Chang, F. K.: Smart layer built-in diagnostics for composite structures, *Proc. of the 4th European Conference on Smart Materials and Structures and the 2nd MIMR Conference*, pp. 777-781, Harrogate, UK, 6-8 July, 1998.

[Chi97]   Chimenti, D. E.: Guided waves in plates and their use in materials characterization, *Applied Mechanics Reviews* 50, pp. 247-284, 1997.

[Cho89]   Cohen, L.: Time-frequency distributions - a review, *Proc. of the IEEE*, 77(7), pp. 941-981, 1989.

[Chu92]   Chui, C. K.: *An Introduction to Wavelets*, Academic Press, San Diego, CA, 1992.

[Com07]   http://surf.to/comet, 2007.

[Doy87]   Doyle, J. F.: Experimentally determining the contact force during the transverse impact of an orthotropic plate, *Journal of Sound and Vibration* 118, pp. 441-448, 1987.

[Egu93]   Egusa, S., Iwasawa, N.: Polling characteristics of PZT/epoxy piezoelectric paints, Ferroelectrics, Vol. 145, pp- 45-60, 1993.

[Gab46]   Gabor, D.: Theory of Communication, *J. IEE London*, 93, 429–457, 1946.

[Gau98]   Gaul, L., Hurlebaus, S.: Identification of the Impact Location using Wavelets, *Mechanical Systems and Signal Processing*, Volume 12, Number 6, pp. 783-795, 1998.

[Giu00]   Giurgiutiu, V. et al.: Active Sensors for Health Monitoring of Ageing Aerospace Structures, *Proc. of SPIE*, Vol. 3985, Newport Beach, CA, 5-9 March, 2000.

[Giu04]   Giurgiutiu, V., Lyshevski, S. E.: *Micromechatronics, Modeling, Analysis, and Design with MATLAB*, CRC Press, 2004.

[Giu07]   Giurgiutiu, V.: *Structural Health Monitoring with Piezoelectric Wafer Active Sensors*, Academic Press, 2007.

[Glo90]   Glossop, N. D., Dubois, W., Tsaw, W. LeBlanc, M., Lymer, J. Measures, R. M., Tennyson, R. C. Optical fiber damage detection for an aircraft composite leading edge, *Composites*, Vol. 21, pp. 71-80, 1990.

[Gun93]   Gunther, M. F., Wang, A., Fogg, B. R., Starr, S. E., Murphy, K. A., Claus, R. O.: Fiber optic impact detection and location system embedded in a composite material, *Proc. SPIE*, Vol. 1798, pp. 262-269, 1993.

[Hal98]   Hallquist, J.: *LS-DYNA - Theoretical Manual*, Livermore Software Technology Corporation, 1998.

[Hal99]   Hall, S. R., Conquest, T. J.: The total data initiative-structural health monitoring, the next generation, *Proc. of the USAF ASIP*, 1999.

[Han92]   Hanselka, H.: *Ein Beitrag zur Characterisierung des Dämpfungsverhaltens polymerer Faserverbundwerkstoffe* [in German], Ph.D dissertation, University of Clausthal, 1992.

[Hay01]   Haywood, J., Staszewski, W. J., Worden, K.: Impact Location in composite structures using smart sensor technology and neural networks, *Proc. of the 3rd International Workshop on Structural Health Monitoring*, pp. 1466-1475, 2001.

[Hof92]   Hoffmann, U.: *Zur Optimierung der Werkstoffdämpfung anisotroper polymerer Hochleistungs-Faserverbundstrukturen* [in German], PhD dissertation, University of Clausthal, 1992.

[Hos03]   Hosten, B., Castaings, M.: Surface impedance matrices to model the propagation in multilayered media, *Ultrasonics*, Vol. 41, pp. 501-507, 2003.

[Hos98]   Hosten, B., Castaings, M., Kundu, T.: Identification of viscoelastic moduli of composite materials from the plate transmission coefficient, *Review of progress in quantitative non-destructive evaluation 17*, pp. 1117-1124, 1998.

[Hur02]   Hurlebaus, S.: *A Contribution to Structural Health Monitoring Using Elastic Waves*, Ph.D dissertation, University of Stuttgart, 2002.

[Iee87]   IEEE, Standard on Piezoelectricity, *ANSI / IEEE*, Std. 176, The Institute of Electrical and Electronics Engineers Inc., New York, 1987.

[Kan56]   Kane, T. R., Mindlin, R. D.: High-frequency Extensional Vibrations of Plates, *Journal of Applied Mechanics*, Vol. 23, pp. 277-283, 1956.

[Kar90]   Karim, M. R., Mal, A. K., Bar-Cohen, Y.: Inversion of leaky Lamb wave data by simplex algorith, *J. Acoust. Soc. Am.* 88(1), pp. 482-491, 1990.

[Khu00]   Khuri-Yakub, B. T., Cheng, C. H., Degertekin, F. L., Ergun, S., Hansen, S., Jin, X. C., Orlakran, O. Silicon micromachined ultrasonic transducer, *Jnp. J. Appl. Phys.*, Vol. 39, pp. 2882-2887.

[Kin90]   Kinra, V. K., Dayal, V.: Dynamic modulus measurement of thin (sub-wavelength specimens, *Dynamic Elastic Modulus Measurements in Materials*, ASTM STP 1045, pp. 18-46, 1990.

[Kis95]   Kishimoto K., Inoue H., Hamada M., Shibuya T.: Time frequency analysis of dispersive waves by means of wavelet transform, *Journal of Applied Mechanics*, 62, pp. 841-846, 1995.

[Kno64]   Knopoff, L.: A matrix method for elastic wave problems, *Bull. Seism. Soc. Am.*, Vol. 54, pp. 54, pp. 431-438, 1964.

[Kro99]   Kropp, W.: *Moving Acoustics - A Library with Auralisation and Visualisation*, Chalmers Vibroacoustic Group, 1999.

[Kub66]   Kubiak, E., Hosek, R., Lichodziejewski, W.: USAF Materials Lab Report, AFML-TR-66-304, 1966.

[Lag73]   Lagasse, P. E.: Higher-order finite element analysis of topographic guides supporting elastic surface waves, *Journal of the Acoustical Society of America*, Vol. 53, pp. 1116-1122, 1973.

[Lag98]   Lagarias, J. C., J. A. Reeds, M. H. Wright, and P. E. Wright: Convergence Properties of the Nelder-Mead Simplex Method in Low Dimensions," *SIAM Journal of Optimization*, Vol. 9, Number 1, pp.112–147, 1998.

[Lak99]   Lakes, R. S.: *Viscoelastic solids*, University of Minnesota, 1999.

[Lam17]   Lamb, H.: On Waves in an Elastic Plate, *Proceedings of the Royal Society of London*, Series A, v.93, n- 651, pp. 293-312, 1917.

[Leh06]   Lehmann, M., Büter, A., Frankenstein, B., Schubert, F., Brunner, B.: Monitoring System for Delamination Detection –Qualification of Structural Health Monitoring (SHM) Systems, *Conference on Damage in Composite Material CDCM*, Stuttgart, September 2006.

[Lia96]   Liang, C., Sun, F., Rogers, C. A. : Electro-mechanical impedance modeling of active material systems, Smart Materials and Structures, 5, pp. 171-186, 1996.

[Lih95]   Lih, S. S., Mal, A. K.: On the Accuracy of Approximate Plate Theories for Wave-Field Calculations in Composite Laminates, *Wave Motion*, 21(1), pp. 17-34, 1995.

[Low95]   Lowe, M. J. S.: Matrix techniques for modeling ultrasonic waves in multilayered media, *IEEE Transactions on Ultrasonics, Ferroelectric, and Frequency Control* 42 pp. 525-542, 1995.

[Mal89]   Mal, A. K., Xu, P. C., Bar-Cohen, Y.: Analysis of leaky Lamb waves in bonded plates, Int. J. Eng. Sci. 27, pp. 779-791, 1989.

[Mar63]   Marquardt, D. W.: An algorithm for least-square estimation of nonlinear parameters, *J. Soc. Ind. Appl. Math.* 11, pp. 431-448, 1963.

[Mea01]   Measures, R. M.: Structural Health Monitoring with Fiber Optic Technology, Academic Press, San Diego, California, 2001.

[Mey06]   Meyendorf, N.: Structural Health Monitoring, Sensor Networks, Intelligent Structures: Attempt of a Prognosis, 2nd *Dresden Airport Seminar on Reliability, Testing, Monitoring of Aerospace Components* , November 2006.

[Min51]   Mindlin, R. D.: Influence of Rotatory Inertia and Shear on Flexural Motions of Isotropic, Elastic Plates, *Journal of Applied Mechanics*, Vol. 18, pp. 31-38, March 1951.

[Min59]   Mindlin, R. D., Medick, M. A.: Extensional Vibrations of Elastic Plates, *Journal of Applied Mechanics*, Vol. 26, pp. 561-569, 1959.

[Nay88]   Nayfeh, A. H., Chimenti, D.E.: Ultrasonic wave reflection from liquid-coupled orthotropic plates with application to fibrous composites, *ASME J. Appl. Mech.* 55, pp. 863-870, 1988.

[Nay89]   Nayfeh, A. H., Chimenti, D. E.: Free wave propagation in plates of general anisotropic media, *Journal of Applied Mechanics*, 56(4), pp. 881-886, 1989.

[Nay91]   Nayfeh, A. H.: The general problem of elastic wave-propagation in multilayered anisotropic media, *Journal of the Acoustical Society of America*, 89(4), pp. 1521-1531, 1991.

[Nay95]   Nayfeh, A. H.: *Wave Propagation in Layered Anisotropic Media*, North-Holland, ISBN: 0-444-89018-1, 1995.

[Nea03]   Neau, G.: *Lamb waves in anisotropic viscoelastic plates, Study of the wave fronts and attenuation*, PhD. dissertation, University of Bordeaux, 2003.

[Nel65]   Nelder, J. A., Mead, R.: A simplex method for function minimization, *Comput. J.* 7, pp. 308-315, 1965.

[Par03]   Park, G. Sohn, H., Farrar, C. F., Inman, D. J.: Overview of Piezoelectric Impedance-Based Health Monitoring and Path Forward, *Shock and Vibration Digest*, Vol. 35(6), pp. 451-463, 2003.

[Par05]   Park, J., Chang, F. K.: System Identification Method for Monitoring Impact Events, *Proc. of SPIE Smart Structures and Materials*, Vol. 5758, 2005.

[Pav97]   Pavlakovic, B. N., Lowe, M. J. S., Alleyne, D. N., Cawley, P.: Disperse: A General Purpose Program for Creating Dispersion Curves, *Review of Progress in QNDE*, Vol. 16, pp. 185-192, 1997.

[Pav98]   Pavlakovic, B. N.: *Leaky Guided Ultrasonic Waves in NDT*, PhD thesis, Imperial College London, 1998.

[Pil88]   Pilarski, A., Rose, J. L.: A transverse-wave ultrasonic oblique incidence technique for interfacial weakness detection in adhesive bonds, *J. Appl. Phys.* 63(2), pp. 300-307, 1988.

[Pol86]   Pollock, A.: Classical Wave Theory in Practical AE Testing, *Progress in AE III, Proceedings of the 8th International AE Symposium,* Japanese Society for Nondestructive Testing, pp. 708-721, November, 1986.

[Pro91]   Prosser, W.: *The propagation characteristics of the plate modes of acoustic emission waves in thin aluminum plates and thin graphite/epoxy composite plates and tubes*, NASA Technical Memorandum 104187, November 1991.

[Qia90]   Qian, Y., Swanson, S. R.: A Comparison of Solution Techniques for Impact response of Composite Plates, *Composite Structures* 14, pp. 177-192, 1990.

[Rag04]   Raghavan A., Cesnik C. E. S.: Modeling of piezoelectric-based Lamb-wave generation and sensing for structural health monitoring, *Proceedings of SPIE*, Vol. 5391, Smart Structures and Materials, Sensors and Smart Structures Technologies for Civil, Mechanical, and Aerospace Systems, Shih-Chi Liu, Editor, pp. 419-430, July 2004.

[Red99]   Reddy, J. N.: *Theory and analysis of elastic plates*, Philadelphia, PA, Taylor and Francis, pp. 446-447, 1999.

[Rei00]   Reid S.R and Zhou, G.: *Impact behaviour of fibre-reinforced composite materials and structures*, CRC Press Cambridge, 2000.

[Rei04]   Reimerdes, H. G.: *Finite Elements in Lightweight Design II*, Lecture notes, University of Aachen RWTH, 2004.

[Rei45]   Reissner, E.: The effect of transverse shear deformation on the bending of elastic plates, *Journal of Applied Mechanics* 12, Transact. ASME 67, pp. 69-77, 1945.

[Roa07]   Roach, D., Rackow, K., DeLong, W., Yepez, S., Reedy, D., White, S.: *Use of Composite Materials, Health Monitoring and Self-Healing Concepts to Refurbish Our Civil and Military Infrastructure*, Sandia Report 2007-5547, September 2007.

[Rog97]   Rogers, C. A., Giurgiutiu, V.: *Electro-Mechanical (E/M) Impedance Technique for Structural Health Monitoring and Non-Destructive Evalua-*

tion, Invention Disclosure No. 97162, University of South Carolina, Office of Technology Transfer, July 1997.

[Roh88]   Rohwer, K.: *Improved Transverse Shear Stiffness for Layered Finite Element*, Research Report No. DVLR-FB 88-32, German Aerospace Center (DLR), 1988.

[Ros99]   Rose, J. L.: *Ultrasonic Waves in Solid Media*, Cambridge University Press, Cambridge, U.K., 1999.

[Sch07]   Schmidt, H. J., Schmidt-Brandecker, B.: Design Benefits in Aeronautics Resulting from Structural Health Monitoring, *Proc. of the 6th International Workshop on Structural Health Monitoring*, Stanford University, Sept. 11-13, 2007.

[Sch95]   Schindler, P. M., May, R. G., Claus, R. O.: Location of impacts on composite panels by embedded fiber optic sensors and neural network processing, *Proc. SPIE*, Vol. 2444, pp. 481-489, 1995.

[Sey01]   Seydel, R., Chang, F. K.: Impact identification of stiffened composite panels: I. system development, *Smart Mater. Struct.*, Vol. 10, pp. 354-369, 2001.

[Sir00]   Sirohi, J., Chopra, I.: Fundamental understanding of piezoelectric strain sensors, *Journal of Intelligent Material Systems and Structures*, Vol. 11, pp. 246-257, April 2000.

[Soh03]   Sohn, H., Farrar, C. et al.: *A Review of Structural Health Monitoring Literature: 1996-2001*, Los Alamos National Laboratory Report, LA-13976-MS, 2003.

[Sri07]   Srivastava, A., Lanza Di Scalea, F.: Generation of Attenuation Curves by the SAFE approach developed at the University of San Diego, *Personal contact and information*, 2007.

[Sta04]   Staszewski, W., Boller, C., Tomlinson, G.: *Health Monitoring of Aerospace Structures, Smart Sensor Technologies and Signal Processing*, Wiley, 2004.

[Sun72]   Sun, C. T., Whitney, J. M.: On Theories for the Dynamic Response of Laminated Plates, *Proceedings AIAA/ASME/SAE 13th Structures, Structural Dynamics and Materials Conference*, AIAA paper no. 72-398, 1972.

[Tho50]   Thomson, W. T.: Transmission of elastic waves through a stratified solid medium, *J. Appl. Phys.*, Vol. 21, pp. 89-93, 1950.

[Udd92]   Udd, E.: Fiber optic sensor, Critical Reviews of Optical Science and Technology, Vol CR44, SPIE Optical Engineering Press, Bellingham, Washington, USA, 1992.

[Ufl48]   Uflyand, Y. S.: The Propagation of Waves in the Transverse Vibrations of Bars and Plates, *Akademiya Nauk SSSR, Prikladnaya Matematica I Mekhanika*, Vol. 12, pp. 287-300, (in Russian), 1948.

[VDI2014] Development of FRP components Analysis, *VDI- Guideline 2014*, Part 3, Sept. 2006.

[Vik67]   Viktorov, I. A.: *Rayleigh and Lamb Waves: Physical Theory and Applications*, Plenum Press, New York, 1967.

[Vis05]   Vishay Micro-Measurements Instruction Bulletion B-127-14, *Strain Gauge Installation with M-Bond 200 Adhesive*, Malvern, PA, 2005.

[Wan04]   Wang, L.: *Elastic Wave Propagation in Composites and Least-squares Damage localization technique*, Master Thesis, University of North Carolina, 2004.

[Wan07]   Wang, L., Yuan, F. G.: Lamb wave propagation in composite laminates using a higher-order plate theory, *Proceedings of the SPIE*, Volume 6531, March 2007.

[Wan90]   Wang, W., Rokhlin, S. I.: Ultrasonic characterization of a thin layer of anodized porous aluminium oxide, *Review of Progress in Quantitative NDE* 9, pp. 1629-1636, 1990.

[Was82]   Washizu, K.: *Variational Methods in Elasticity and Plasticity*, 3rd Edition, Pergamon Press, Oxford, 1982.

[Whi70]   Whitney, J. M., Pagano, N. J.: Shear Deformation in Heterogeneous Anisotropic Plates, *Journal of Applied Mechanics, Trans. ASME*, Vol. 37, pp. 1031-1036, 1970.

[Whi73]   Whitney, J. M., Sun, C. T.: A Higher Order Theory for Extensional Motion of Laminated Composites, *Journal of Sound and Vibration*, Vol. 30, No. 1, pp. 85-97, 1973.

[Wik07]   http://en.wikipedia.org/wiki/Aloha_Flight_243, 2007.
[Wik07a]   http://en.wikipedia.org/wiki/China_Airlines_Flight_611, 2007.
[Wil01]   Wilcox, P. D., Lowe, M. J. S., Cawley, P.: The effect of dispersion on long-range inspection using ultrasonic guided waves, *NDT&E International*, Vol. 34, pp. 1-9, 2001.

[Wil06]   Wilcox, P. D., Croxford, A. J., Konstantinidis, G., Drinkwater, B.: Guided Waves Structural Health Monitoring, *2nd Dresden Airport Seminar*, 2006.

[Wit87]   Wittrick W. H.: Analytical, three dimensional elasticity solutions to some plate problems and some observations on Mindlin's plate theory, *Int J Solids Struct*, Vol. 23, pp. 441–64, 1987.

[Wor00]   Worden, K., Staszewski, W. J.: Impact location and quantification on a composite panel using neural networks and a genetic algorithm, *Strain*, Vol. 36, pp. 61-70, 2000.

[Wor57]   Worlton, D. C.: Ultrasonic Testing with Lamb waves, *Non-destructive Testing*, Vol. 15, pp. 218-222, 1957.

[Yan05]   Yang, S., Yuan, F. G.: Transient wave propagation of isotropic plates using a higher-order plate theory, *International Journal of Solids and Structures*, 42, pp. 4115-4153, 2005.

[Yos95]   Yoshikawa, S., Selvaraj, U., Moses, P., Withams, J., Meyer, R., Shrout, T.: Pb(Zr,Ti)O3 [PZT] Fibers-Fabrication and Measurement Methods, *Journal of Intelligent Material Systems and Structures*, Vol. 6, No. 2, pp. 152-158, 1995.

[Yu06]   Yu, L.: *In-Situ Structural Health Monitoring with Piezoelectric Wafer Active Sensor Guided-Wave Phased Arrays*, Ph.D dissertation, University of South Carolina, 2006.

[Zho98]   Zhou, G.: The use of experimentally-determined impact force as a damage measure in impact damage resistance and tolerance of composite structures, *Compos. Struct*, 42, pp. 375-382, 1998.

[Zin94]   Zinoviev, P. A., Ermakov, Y. N.: *Energy dissipation in Composite Materials*, ISBN 1-56676-082-8, 1994.

12 Appendix

12.1 Excitability results

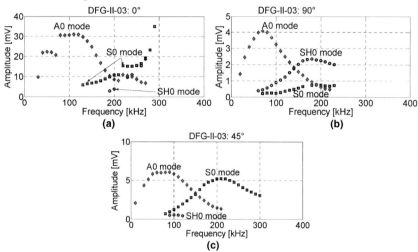

Fig. 12.1: Excitability of Lamb waves in the CFRP laminate "*DFG-II-03*":

(a) in 0° direction @ PWAS#2; (b) in 90° direction @ PWAS#9; (c) in 45° direction @ PWAS#5

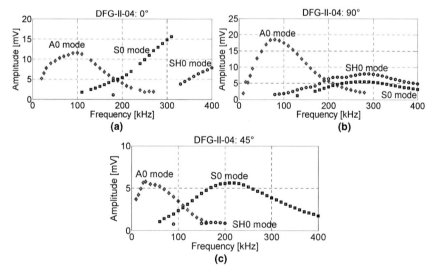

Fig. 12.2: Excitability of Lamb waves in the CFRP laminate "*DFG-II-04*":

(a) in 0° direction @ PWAS#2; (b) in 90° direction @ PWAS#8; (c) in 45° direction @ PWAS#5

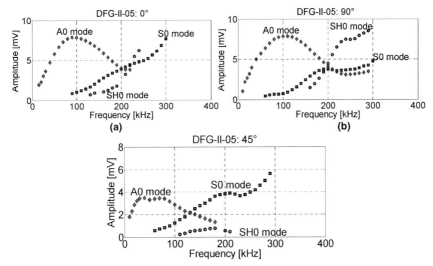

Fig. 12.3: Excitability of Lamb waves in the CFRP laminate "*DFG-II-05*":
(a) in 0° direction @ PWAS#2; (b) in 90° direction @ PWAS#8; (c) in 45° direction @ PWAS#5

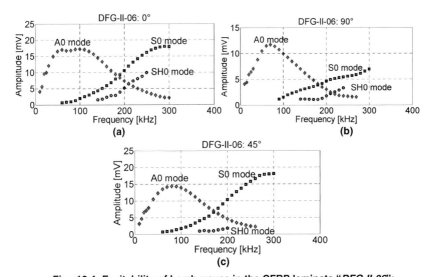

Fig. 12.4: Excitability of Lamb waves in the CFRP laminate "*DFG-II-06*":
(a) in 0° direction @ PWAS#2; (b) in 90° direction @ PWAS#8; (c) in 45° direction @ PWAS#5

12.2 Dispersion diagrams

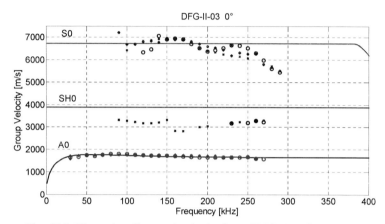

Fig. 12.5: Dispersion diagram for the laminate "DFG-II-03" in 0° direction

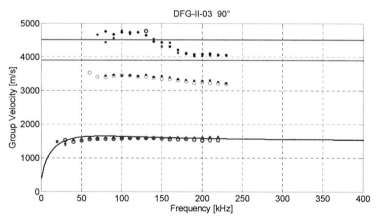

Fig. 12.6: Dispersion diagram for the laminate "DFG-II-03" in 90° direction

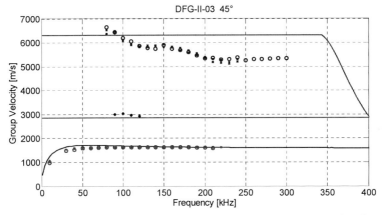

Fig. 12.7: Dispersion diagram for the laminate "DFG-II-03" in 45° direction

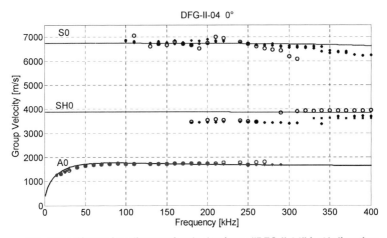

Fig. 12.8: Dispersion diagram for the laminate "DFG-II-04" in 0° direction

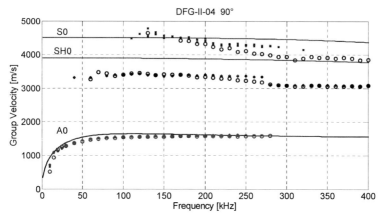

Fig. 12.9: Dispersion diagram for the laminate "DFG-II-04" in 90° direction

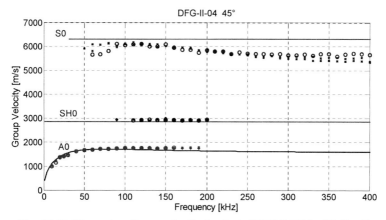

Fig. 12.10: Dispersion diagram for the laminate "DFG-II-04" in 45° direction

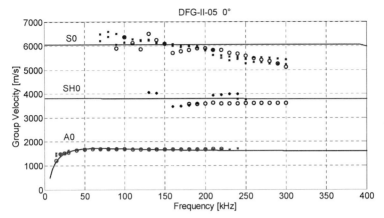

Fig. 12.11: Dispersion diagram for the laminate "DFG-II-05" in 0° direction

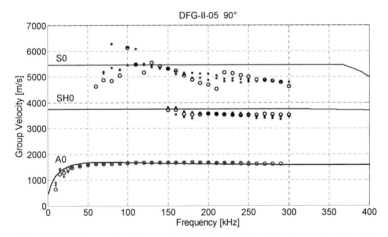

Fig. 12.12: Dispersion diagram for the laminate "DFG-II-05" in 90° direction

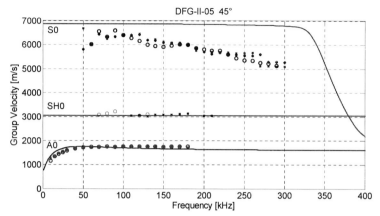

Fig. 12.13: Dispersion diagram for the laminate "DFG-II-05" in 45° direction

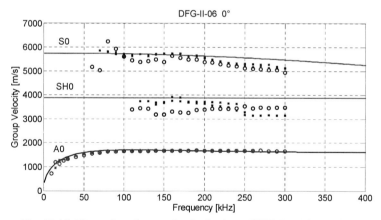

Fig. 12.14: Dispersion diagram for the laminate "DFG-II-06" in 0° direction

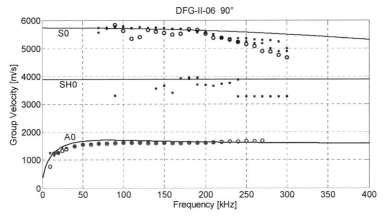

Fig. 12.15: Dispersion diagram for the laminate "DFG-II-06" in 90° direction

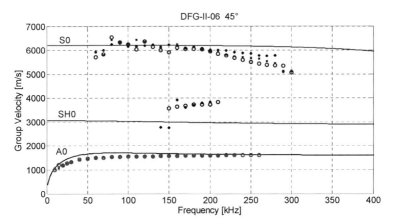

Fig. 12.16: Dispersion diagram for the laminate "DFG-II-06" in 45° direction

12.3 Attenuation dispersion diagrams

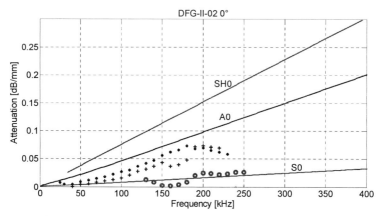

Fig. 12.17: Attenuation dispersion diagram for the laminate "DFG-II-02" in 0° direction

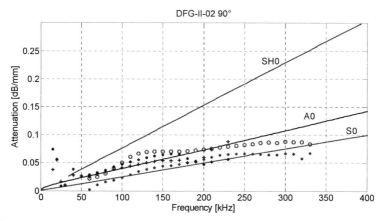

Fig. 12.18: Attenuation dispersion diagram for the laminate "DFG-II-02" in 90° direction

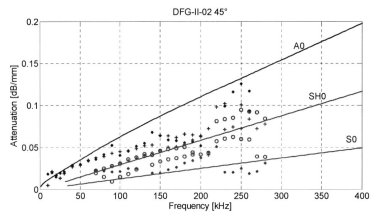

Fig. 12.19: Attenuation dispersion diagram for the laminate "DFG-II-02" in 45° direction

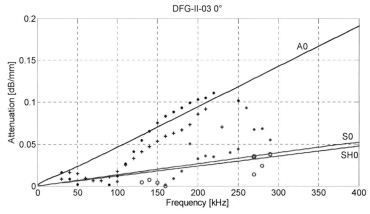

Fig. 12.20: Attenuation dispersion diagram for the laminate "DFG-II-03" in 0° direction

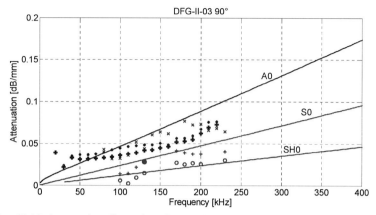

Fig. 12.21: Attenuation dispersion diagram for the laminate "DFG-II-03" in 90° direction

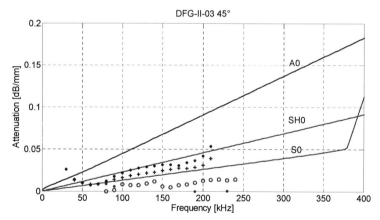

Fig. 12.22: Attenuation dispersion diagram for the laminate "DFG-II-03" in 45° direction

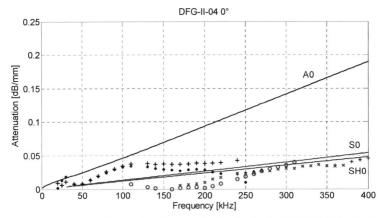

Fig. 12.23: Attenuation dispersion diagram for the laminate "DFG-II-04" in 0° direction

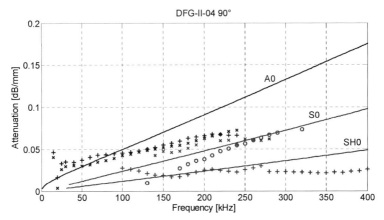

Fig. 12.24: Attenuation dispersion diagram for the laminate "DFG-II-04" in 90° direction

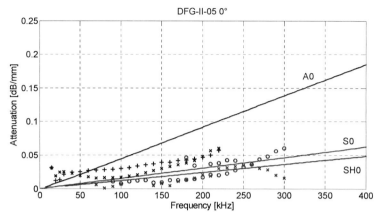

Fig. 12.25: Attenuation dispersion diagram for the laminate "DFG-II-05" in 0° direction

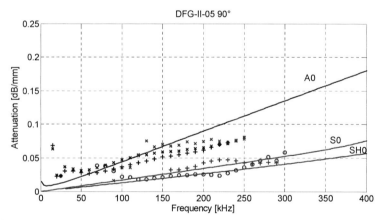

Fig. 12.26: Attenuation dispersion diagram for the laminate "DFG-II-05" in 90° direction

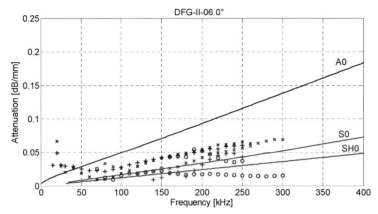

Fig. 12.27: Attenuation dispersion diagram for the laminate "DFG-II-06" in 0° direction

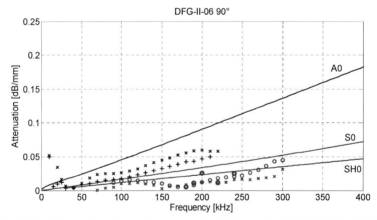

Fig. 12.28: Attenuation dispersion diagram for the laminate "DFG-II-06" in 90° direction

Bisher erschienene Bände der Reihe

Science-Report aus dem Faserinstitut Bremen

ISSN 1611-3861

10   Hendrik Mainka   Lignin as an Alternative Precursor for a Sustainable and Cost-Effective Carbon Fiber for the Automotive Industry
ISBN 978-3-8325-3972-6   43.00 EUR

Alle erschienenen Bücher können unter der angegebenen ISBN-Nummer direkt online (http://www.logos-verlag.de) oder per Fax (030 - 42 85 10 92) beim Logos Verlag Berlin bestellt werden.